D0064815

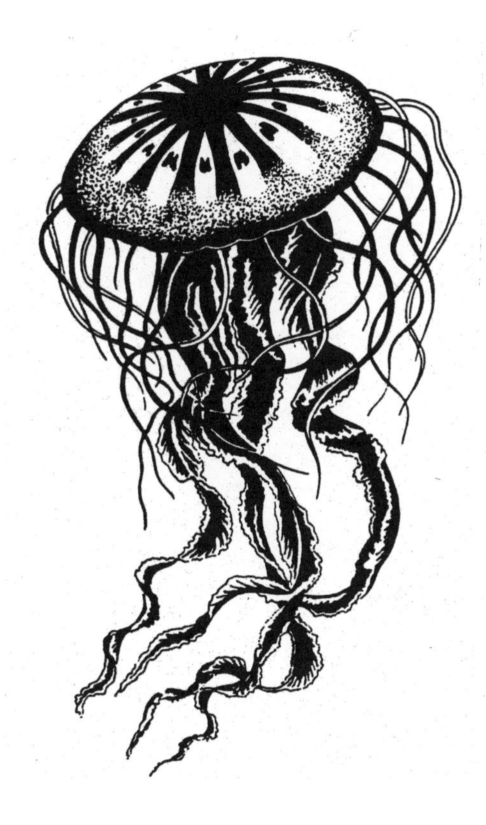

Published by
Bloomsbury USA, New York
Bloomsbury is a trademark of Bloomsbury Publishing Plc

All papers used by Bloomsbury USA are natural, recyclable products made from wood grown in well-managed forests. The manufacturing processes conform to the environmental regulations of the country of origin.

Library of Congress Cataloging-in-Publication Data
has been applied for.

ISBN-13: 978-0-8027-1538-8

First U.S. edition 2006

7 9 10 8

Designed and typeset by
Wooden Books Ltd, Glastonbury, UK

Printed in the U.S.A. by Worzalla,
Stevens Point, Wisconsin

SYMMETRY

THE ORDERING PRINCIPLE

written and illustrated by

David Wade

B L O O M S B U R Y
NEW YORK • LONDON • NEW DELHI • SYDNEY

For Emile Boulanger

All pictures by the author, except for Japanese "pine-bark" pattern on page 39, reproduced from Japanese Patterns, by Jeanne Allen, with kind permission of Chronicle Books; and the Portrait of Emmy Noether on page 47 by Jesse Wade.

For further reading, try "Symmetry and the Beautiful Universe," by Leon Lederman and Christopher Hill, Mario Livio's "The Equation that couldn't be Solved," or "Symmetry, a Unifying Concept" by Istvan and Magdolna Hargittai.

"Let proportions be found not only in numbers and measures, but also in sounds, weights, times and positions, and whatever force there is" - Leonardo da Vinci.

Above - Da Vinci's conjecture that the total cross-sectional area of a tree remains the same at all branching levels; a balance illustrates the hidden symmetry of force equal to mass x distance. Overleaf a sample of nature's infinite symmetrical variety; Ernst Haekel's drawings of various species of diatoms.

CONTENTS

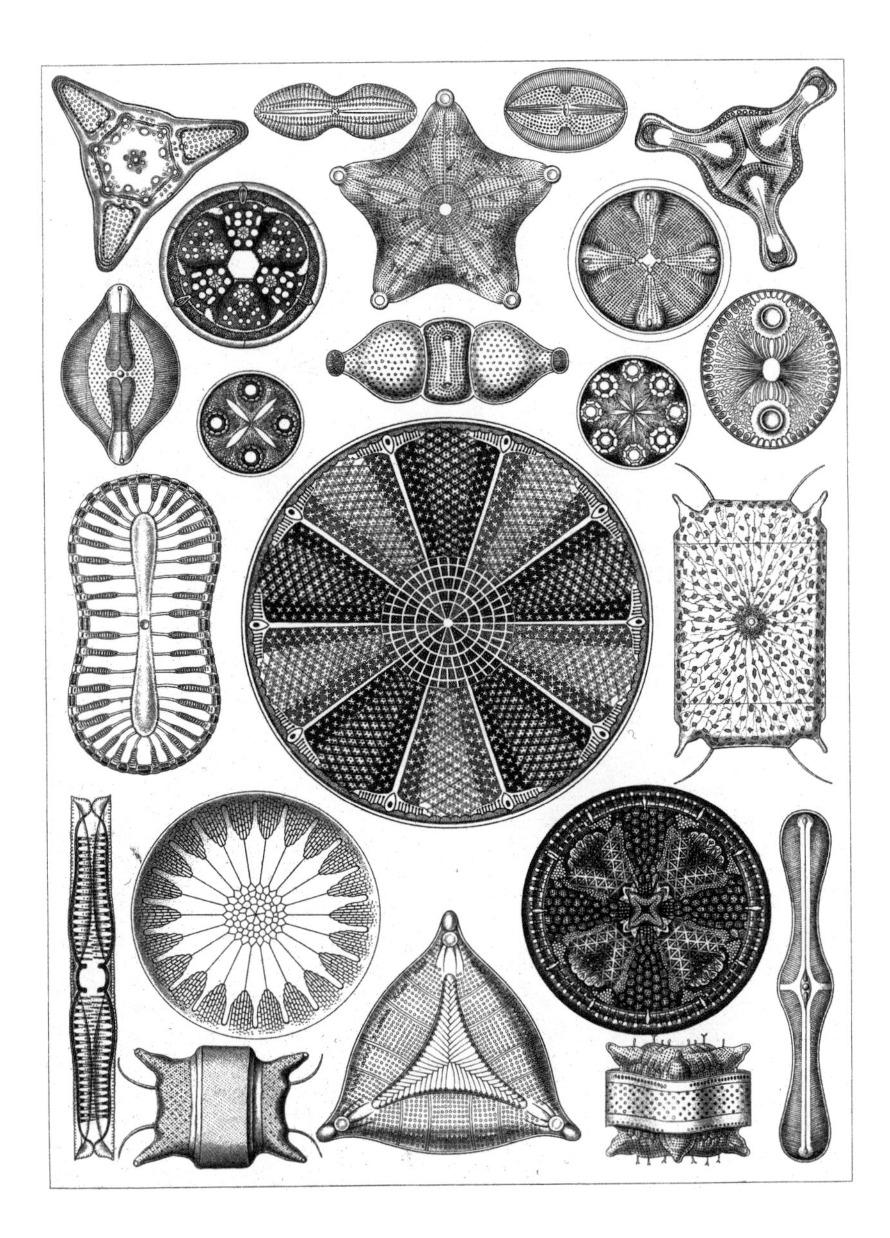

INTRODUCTION

Symmetry has a very wide appeal; it is of as much interest to mathematicians as it is to artists, and is as relevant to physics as it is to architecture. In fact, many other disciplines lay their own claims on the subject, each having their own ideas of what symmetry is, or should be. Clearly, whatever approach is taken, we are dealing here with a universal principle, however, in our day-to-day experience conspicuous symmetries are comparatively rare and most are far from obvious. So what is symmetry? Are there general terms for it? Can it, indeed, be clearly defined at all?

On investigation, it soon becomes clear that the whole field is hedged about with paradox. To begin with, any notion of symmetry is completely entangled with that of asymmetry; we can scarcely conceive of the former without invoking thoughts of the latter (as with the related concepts of order and disorder) - and there are other dualities. Symmetry precepts are always involved with categorization, with classification and observed regularities; in short, with limits. But in itself symmetry is unlimited; there is nowhere that its principles do not penetrate. In addition, symmetry principles are characterized by a quietude, a stillness that is somehow beyond the bustling world; yet, in one way or another, they are almost always involved with transformation, or disturbance, or movement.

The more deeply one investigates this subject the more apparent it becomes that this is at the same time one of the most mundane and extensive areas of study—but that, in the final analysis, it remains one of the most mysterious.

ARRAYS
the regular disposition of elements

When it comes to understanding just what the common factors are among the many and various aspects of symmetry, the notions of *congruence* and *periodicity* take us a long way. Most symmetries present these aspects in one form or another, and the absence of one or the other usually leads to a reduction, or even the lack of symmetry.

For instance: two like objects, in no particular relation with each other, are merely similar (since although they may be congruent they are not arranged in any order) (*1, opposite*). The addition of a third object allows a degree of regularity to come into play, creating the basis of a recognizable pattern (*2*).

So, in its simplest form, symmetry is expressed as a regularly repeating figure along a line (*below*), a series that may easily be extended into an *array* (*3*). Obviously, simple arrangements of this kind could in theory be indefinitely extended, but symmetry will be maintained just so long as both the repeating element and the spacing remain consistent.

We can recognize array symmetries in many natural formations, from the familiar rows of kernels in sweet corn (*4*), to the patterns of scales in fish and reptiles (*5*). And of course such regular arrangements feature in a great deal of human art and artefacts—as in the decorated shaman's cloak opposite (*6*). Naturally, there are often functional as well as aesthetic criteria operating in the formation of arrays, which is evident in the sort of patterns created by brickwork and roof-tiles (*7,8*).

1. *Mere similarity.*

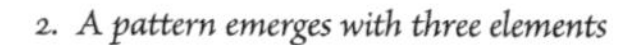

2. *A pattern emerges with three elements*

3. *Symmetrical arrays involve regular spacing. In essence, all symmetries are based on "invariance" or "self-coincidence." In geometric symmetry, the imagined movement that is necessary to achieve this, whether it involves simple repetition, reflection, or rotation (see over), is known as an isometry (see Appendix).*

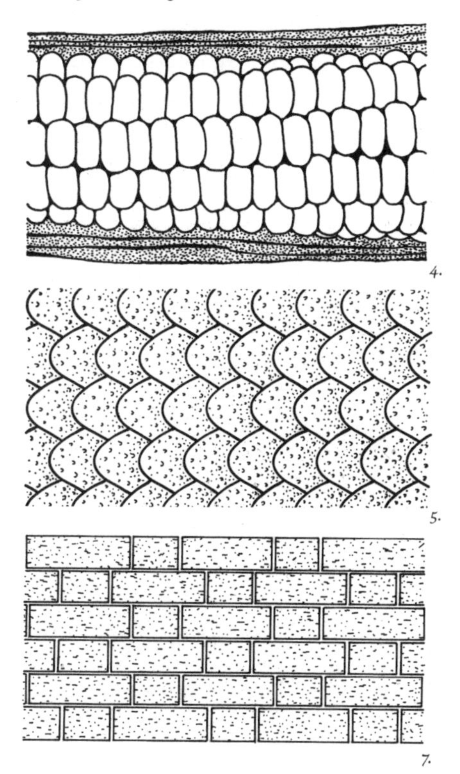

4.

5.

7.

6.

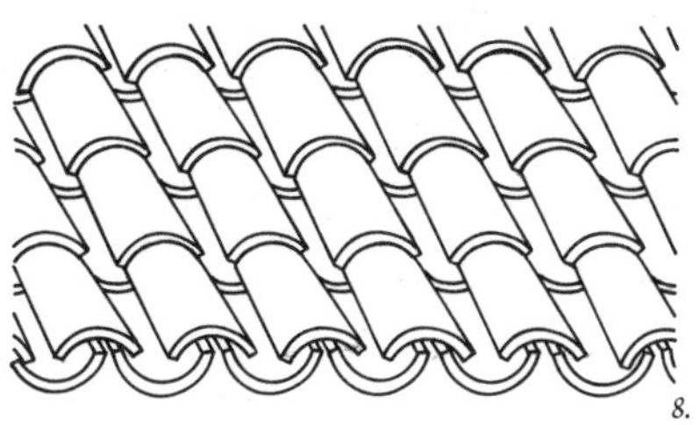

8.

ROTATIONS AND REFLECTION

point symmetries

There are two further basic expressions of symmetry, namely *rotation* and *reflection*. Each of these forms of symmetry relies on the notion of *congruence*, that is to say, a general correspondence between each part of an element, however expressed (*below*). In simple rotational symmetry the component parts are laid out, at regular intervals, around a central point (*1-4*).

Because the elements in these symmetries are simple unreversed replicas of each other, they are described as being directly congruent. In reflection symmetry, by contrast, the reversed elements are arranged about a mirror line, and so are oppositely congruent (*5,6*). Because the central point or line remains fixed in reflections and rotations, these are collectively known as *point symmetries*.

In its most basic form, rotational symmetry involves just two components arranged around a center. Ordinary playing cards are of this kind - any cut through the center of a card results in two identical halves. The triskelion symbol consists of three rotated parts; a swastika of four, and so on – with no upward limit to the number, other than the amount of repeats that can be arranged around a given center.

Rotational and reflection symmetries can also be combined, in which case the lines of reflection intersect at a central point of rotation. Figures and objects of this kind are described as having *dihedral* symmetry (*7*).

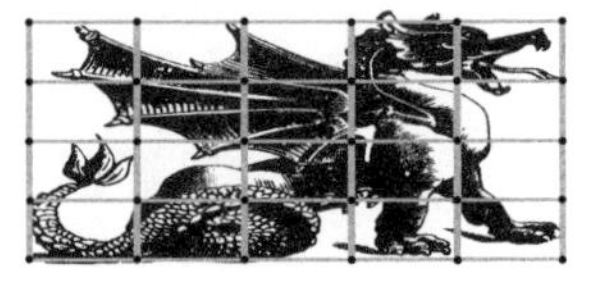

1. The simplest form of rotation around a center, using just two elements

2. Playing-cards are probably the most familiar example of 2-rotation symmetry, demonstrating a self-coincidence of 180° (note that there is no reflection here)

3. Rotational symmetry may involve any number of elements

4. Motifs using 3-, 4- and 5- rotational symmetry , with self-coincidences of 120°, 90° and 72° respectively

5. Reflection about a line

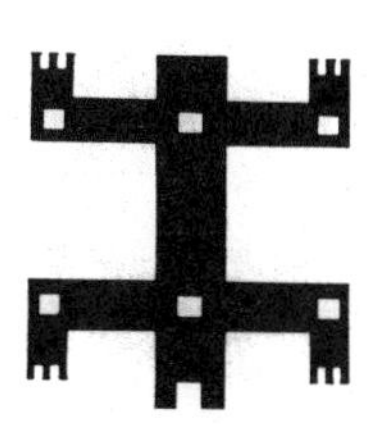

6. Motifs with only reflection symmetry are among the most common

7. Dihedral symmetry

8. Motifs demonstratng dihedral symmetry, combining reflection and rotation

GEOMETRIC SELF-SIMILARITY

gnomons and other self-similar figures

Symmetry is an invariable characteristic of both growth and form, whether in simple or complex, living or non-living, systems.

The *gnomon* demonstrates one of the simplest examples of geometrical growth (*see below*). The principle is this: when a gnomon is added to another figure that figure is enlarged but retains its general shape—and this can be carried on indefinitely. This is essentially what happens in the elaborate forms created by shells and horns, where new growth is added to dead tissue.

Dilation symmetries also produce figures that are geometrically similar to an original. These derive from the enlargement (or reduction) of a form by way of lines radiating from a center. Dilation symmetries, which may extend from the infinitely small to the infinitely large, can use any angle from a center (*1*), or any regular division of the circle (*2*), or its entirety (*3*).

Dilation may also be linked to rotation, producing *continuous* symmetries that can give rise to equiangular spirals (*4*) (of which more later), or discontinuous symmetries (*5*), (in which case the increments are not necessarily a sub-multiple of a complete turn). Dilation symmetries also occur in three-dimensional space. As can be seen, spiral symmetries are intimately connected with the movements of rotation and dilation, and tend to emerge whenever these are combined.

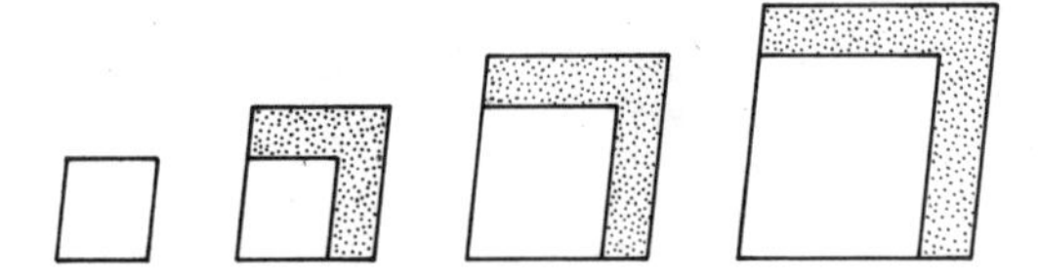

1. Dilation symmetries involve regular increase (or decrease)

2. Point-centered dilation

3. Dilation over 360°

4. Dilation combined with rotation

5. Discontinuous rotated dilation

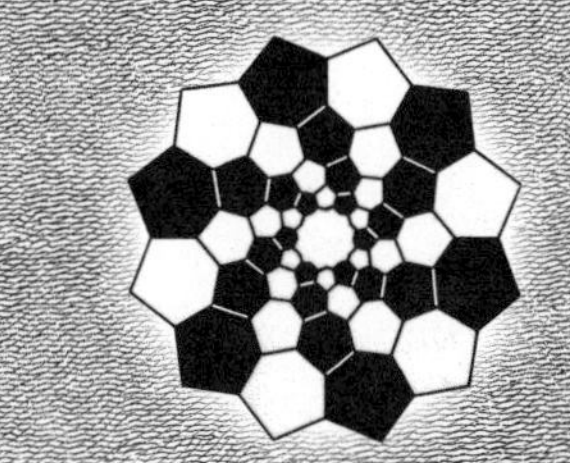

6. Similarity symmetries arising from the regular arrangement of figures

RADIAL
centered symmetries

Radial symmetries are probably the most familiar of all regular arrangements. Being finite, they belong to the broad category of point-group symmetries—and they come in three distinct forms.

In two dimensions they are centered on a point in the plane, showing rotational symmetry, with any number of regular divisions of the circle; reflection is also frequently incorporated, creating dihedral symmetries (*1*). Many flowers show this arrangement, and of course centered, radial motifs appear in the decorative art of practically every culture.

In three dimensions, radial symmetries are either centered on a point in space, where each path fans out from the center to every outlying point (as in an explosion) (*2*); or they have a polar axis of rotation, typically cylindrical or conical (*3*). These last are the characteristic symmetries of plants.

The great majority of flowers have petal arrangements using a number taken from the Fibonacci series, i.e. 3, 5, 8, 13, 21 etc. (*more on this magical sequence on page 30*). The celebrated symmetry of snow-crystals, by contrast, is always six-pointed.

As well as being a favored symmetry of decorative motifs, planar-radial symmetry is also the most useful configuration for any device involving rotary motion—particularly the wheel in its various manifestations.

Radial symmetries of all kinds, being finite, belong to the category of point-group symmetries

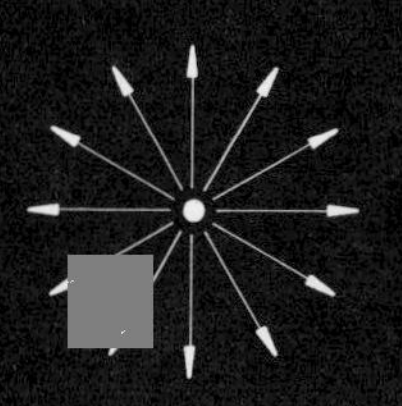

1. 2-d radial symmetry

2. 3-d radial symmetry

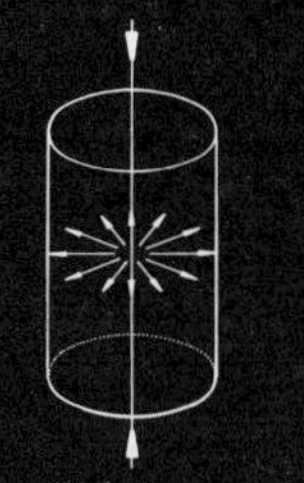

3. Radial symmetries around a polar axis

SECTIONS AND SKELETONS

internal symmetries of plants and animals

The great majority of plants express radial symmetry in one form or another. In fact the great divide between the kingdoms of plants and animals is reflected in their dominant symmetries. Because plants are usually fixed and nonmotile they tend to be radial, whereas the majority of animals move of their own volition and as a result are *bi-lateral,* or, more accurately, *dorsiventral* (*see page 18*).

The trunks and branches of trees usually indicate a radial arrangement in transverse cross-section, and the same is true of roots and vertical stems in general (*1*). Most regular (actinomorphic) flowers have a radial symmetry, as do many inflorescences (*2*). Placentation, too, is invariably arranged on a symmetrical plan (*below*). Mushrooms, mosses and the tubular leaves of rushes also adopt this symmetry.

Sessile animals, i.e. those which are attached and unable to move under their own power, usually have a plant-like, radial symmetry. The predominant number of these are marine creatures, such as sea-anemones and sea-urchins (*3*). Starfish and star corals are likewise center-structured.

The jewel-like skeletons of the marine Protozoa (which include the Radiolaria and Foraminifera), which are found in such profusion in the seas that they account for up to 30% of ocean sediments, also tend to adopt radial symmetries in their body form (*4*).

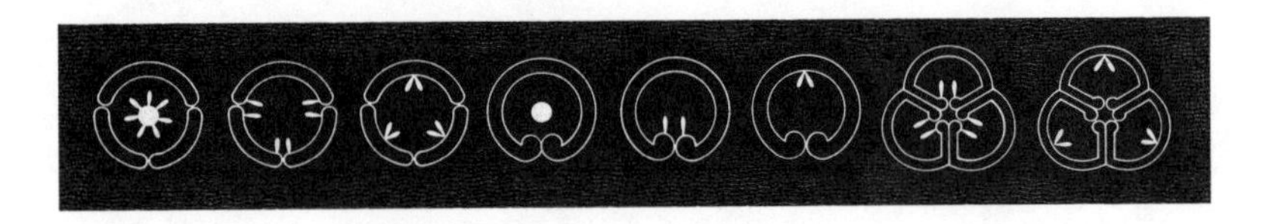

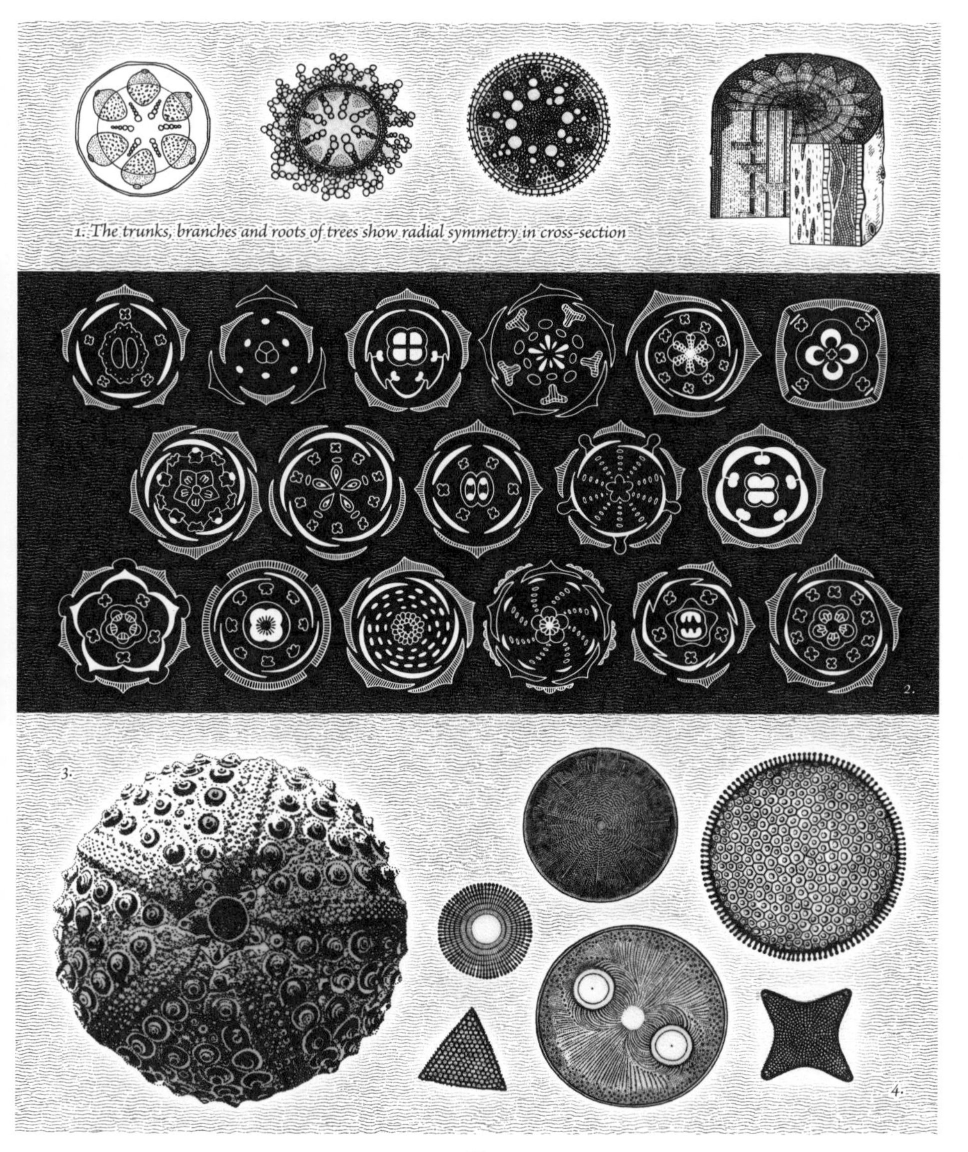

1. The trunks, branches and roots of trees show radial symmetry in cross-section

SPHERICAL

the perfect 3-dimensional symmetry

Just as the circle is the perfect figure in two dimensions, an ideal sphere is a perfect, radially symmetric, 3-D body. Both were known as such by the Ancient Greeks, and were considered divine (the philosopher Xenophanes went so far as to replace the old pantheon of Gods with a single deity, which he assumed to be spherical). Pythagoras was the first to teach that the Earth itself was spherical in shape; more recent cosmologists have suggested that the entire, expanding cosmos has the overall symmetry of a sphere. Interestingly, this shape appears at the very opposite extremes of scale—stars, planets, moons, the Oort cloud and the globular clusters of galaxies are spherical (*1*), and so are small water droplets. Each owe their symmetrical regularity to the fact that they are shaped by a single dominant force; the latter to surface tension, all of the former to gravity (which itself is spherically symmetric).

The action of surface tension is also responsible for the spherical shape of a host of microscopic creatures (*2*). These tend to be virtually fluid in composition and have to maintain an internal pressure that is in balance with that of their surrounding medium. In fact most spherical creatures tend to be very small (where the distorting effects of gravity are minimized), and to live in water. The great majority of these have little or no motivity. In practical terms a sphere represents the smallest surface area for a given volume, which is why so many fruit (*3*) and eggs (*5*) are this shape. Since it minimizes surface area, and presents the same profile on every side, the sphere also offers a natural defense against predation. Hence the evolved response in those species which, whilst not spherical to begin with, roll themselves up into balls when attacked (*4*).

1.
2.
3.
4.
5.

SYMMETRIES IN 3-D
spatial isometries

Just as the sphere is the three-dimensional equivalent of the perfect symmetry of the two-dimensional circle, the *transformations* of figures in space correspond with that of the regular division of the plane that we saw earlier, and similar isometric principles are involved (*1-6*).

If we look to the ways in which space can be symmetrically partitioned, the most elementary divisions follow from the regular plane-filling figures. So, just as the equilateral triangle, square and hexagon fill two dimensions, the prisms based on these will completely fill space (*7*). When it comes to space-fillers that are regular in all directions the options are rather less obvious, but include the cube, the truncated octahedron (*5*), the cubeoctahedral system (*8*) and the rhombic dodecahedron (*9*). The three spherical symmetrical systems (*10*) have a particular bearing on the regular solid figures.

Interestingly, among the huge variety of regular figures, nature consistently chooses one family above all others, namely, the pentagonal dodecahedra. These shapes, made up of hexagons and pentagons, are adopted by forms as diverse as the Fullerene molecule, soot-particles, radiolaria and viruses (*below*). The intriguing aspect of these shapes, and perhaps the key to their usefulness in nature, is that while hexagons themselves cannot enclose space, any number can be enabled to do so with the addition of just twelve pentagons.

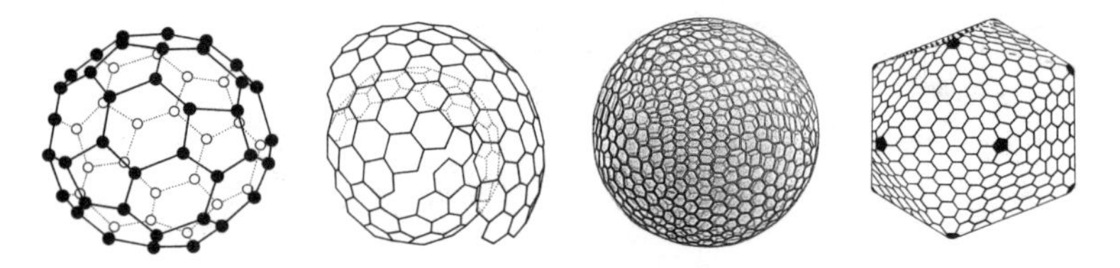

1. 3-D symmetry along a line

2. 3-D rotation about an axis

3. 3-D reflection about a mirror-plane

4. 3-D point-group symmetry

5. 3-D space-group symmetry

6. 3-D dilation symmetry

7. Space-filling prisms

8. Cubeoctahedral system

9. Rhombic dodecahedron

10. The three spherical systems of symmetry: tetrahedral, octahedral, and icosahedral.

STACKING AND PACKING

fruit, froth, foams and other space-fillers

Finding the easiest and most efficient way of stacking a pile of oranges in a given area is one of those deceptively simple-sounding tasks that have far-reaching mathematical ramifications. The problem is simple enough to begin with. The most obvious ways of packing spherical objects together are the triangular and square arrangements, (*1-3*); these configurations obviously relate to the regular division of the plane (*see Appendix*). Having laid out the fruit in either of these patterns it is difficult to stack a second layer other than in the interstices formed by the first. They tend to fall, literally, into a pattern of minimum energy. There are three distinct cubic arrangements (*4,5,6*), but the face-centered assembly has been shown to be the most efficient—although a final proof came only 400 years after Kepler first proposed it.

In many other circumstances, however, three-way junctions of 120° provide the most economical systems. Bee-cells, of course, are the classic example. They use the minimum amount of wax to create storage containers for their honey (*7*). It is also the case that small groups of soap bubbles with free boundaries pull themselves into this efficient angular formation, known as the Plateau border (*8*).

When it comes to larger clusters of soap-bubbles, however, an entirely different magic angle is involved, namely 109° 28' 16". In any froth or elastic foam (*9*) the interior surfaces tend to meet at this angle - which is exactly that formed by a line from the center to the corner of a tetrahedron (*10*). Interestingly, as a solid figure, the tetrahedron by itself will not completely fill space—although it will in combination with the octahedron.

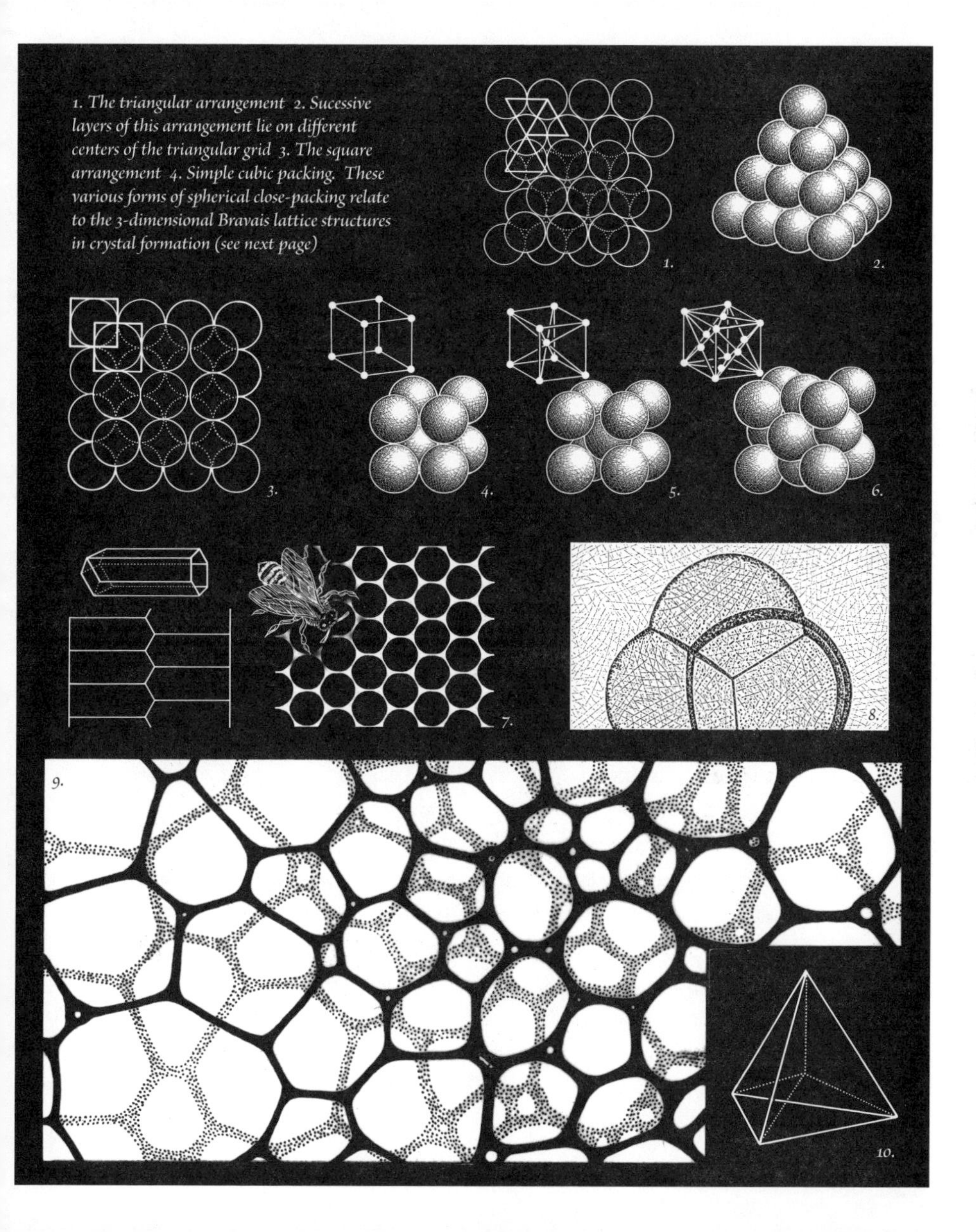

1. The triangular arrangement 2. Sucessive layers of this arrangement lie on different centers of the triangular grid 3. The square arrangement 4. Simple cubic packing. These various forms of spherical close-packing relate to the 3-dimensional Bravais lattice structures in crystal formation (see next page)

The Crystalline World

the stronghold of symmetrical order

Of all natural objects, well-formed crystals make the closest approximation to the mathematical purity of the regular solids (they can indeed assume some of these shapes, but not all). However, the fascinating, pristine beauty of specimen crystals is simply an externalization of an even more impressive internal structure. In fact the crystalline state, with its constituent molecules lined up in tens, or even hundreds of millions, of obedient, identical molecules, is a realm of almost inconceivable orderliness.

Crystals of different substances adopt a wide range of different and characteristic forms, but their regularities are based on the unit-cell arrangements of one or other of just fourteen lattice structures (*below*). These Bravais lattices, the equivalent of 2-D graphs, enable the component molecules to repeat indefinitely in three different spatial directions, much like the "repeat" of a wallpaper pattern.

The early scientific investigation of crystals was primarily concerned with classification, primarily in terms of the symmetries involved. By the mid-19th century crystals had been placed in thirty-two distinct classes, and by the end of that century all 230 possible space-groups had been listed by the Russian crystallographer Federov.

The discovery of X-ray diffraction in the early 20th century, however, completely transformed the science. Systematic analysis of the symmetrical patterns thrown onto a photographic plate by this method revealed for the first time the extraordinary internal world of crystals.

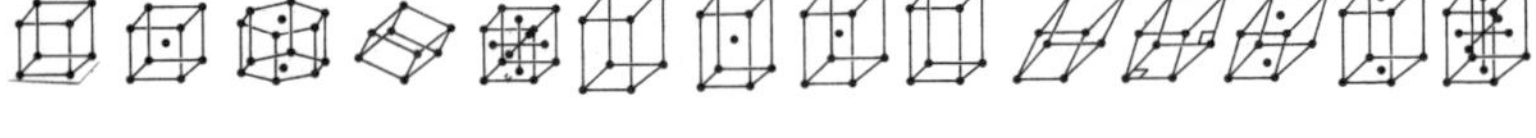

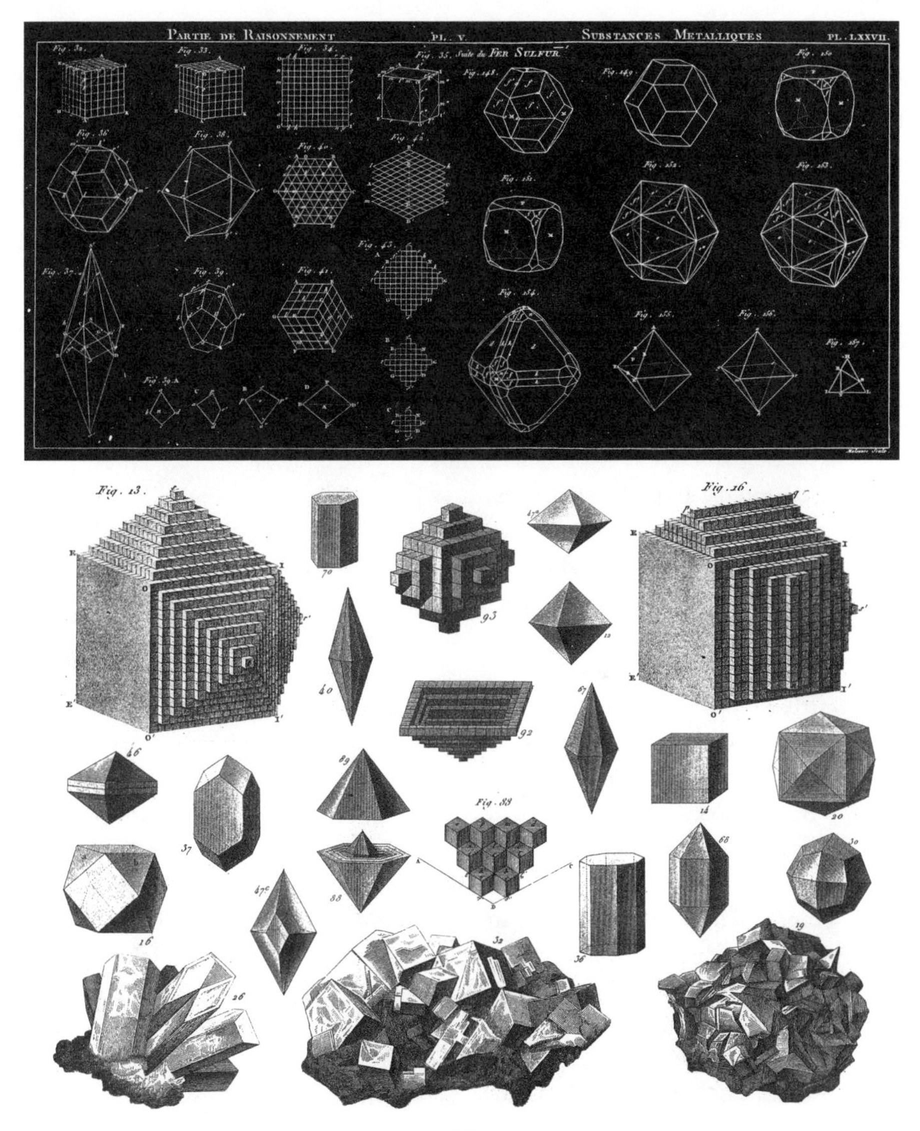
PARTIE DE RAISONNEMENT
PL. V.
SUBSTANCES METALLIQUES
PL. LXXVII.
Fig. 35. Suite du FER SULFURÉ
Fig. 13.
Fig. 16.
Fig. 88

BASIC STUFF

symmetries at the heart of matter

Toward the end of the 19th century the pioneering physicist Pierre Curie stated what he felt to be a universal principle of physics, to the effect that symmetric causes will necessarily lead to equally symmetric effects. Now, as a general principle he was quite wrong, for symmetries are not always linked in the way that he implied. But his intuition of symmetric continuity is certainly true at the more basic levels of matter. The highly ordered world of the crystalline state, exposed to view by X-ray crystallography (*1*), is entirely determined by the underlying symmetries of the atomic and sub-atomic realms.

Mendeleev's Periodic Table, which placed the elements into a rational series, was one of the great milestones of 19th-century, classical physics. But early on in the 20th century it became clear that the properties of the elements were, in fact, reflecting regularities within the internal structures of their component atoms. As atomic theories developed further it became apparent that all chemical properties derived from the numbers of protons and electrons in their respective atomic structures, allowing them to group in orderly molecular arrangements (*2*).

By the 1960s it was realized that although the "orbiting" electrons (*3*) were indeed fundamental particles, the protons and neutrons of the nucleus (*4*) were made up of yet smaller components—hadrons and leptons. The hadrons, in turn, are combinations of quarks which come together in the beautiful symmetries of the famous "8-fold way" (*5*).

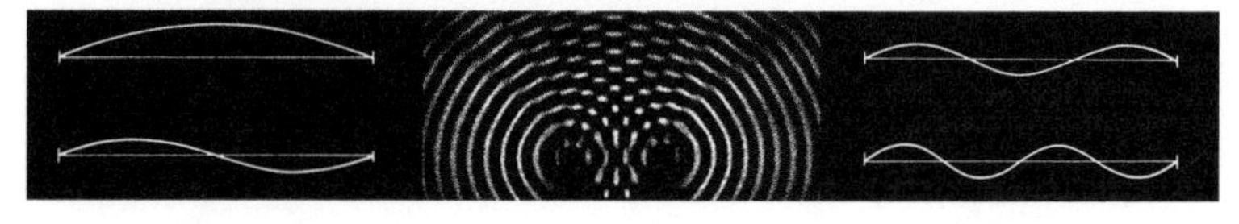

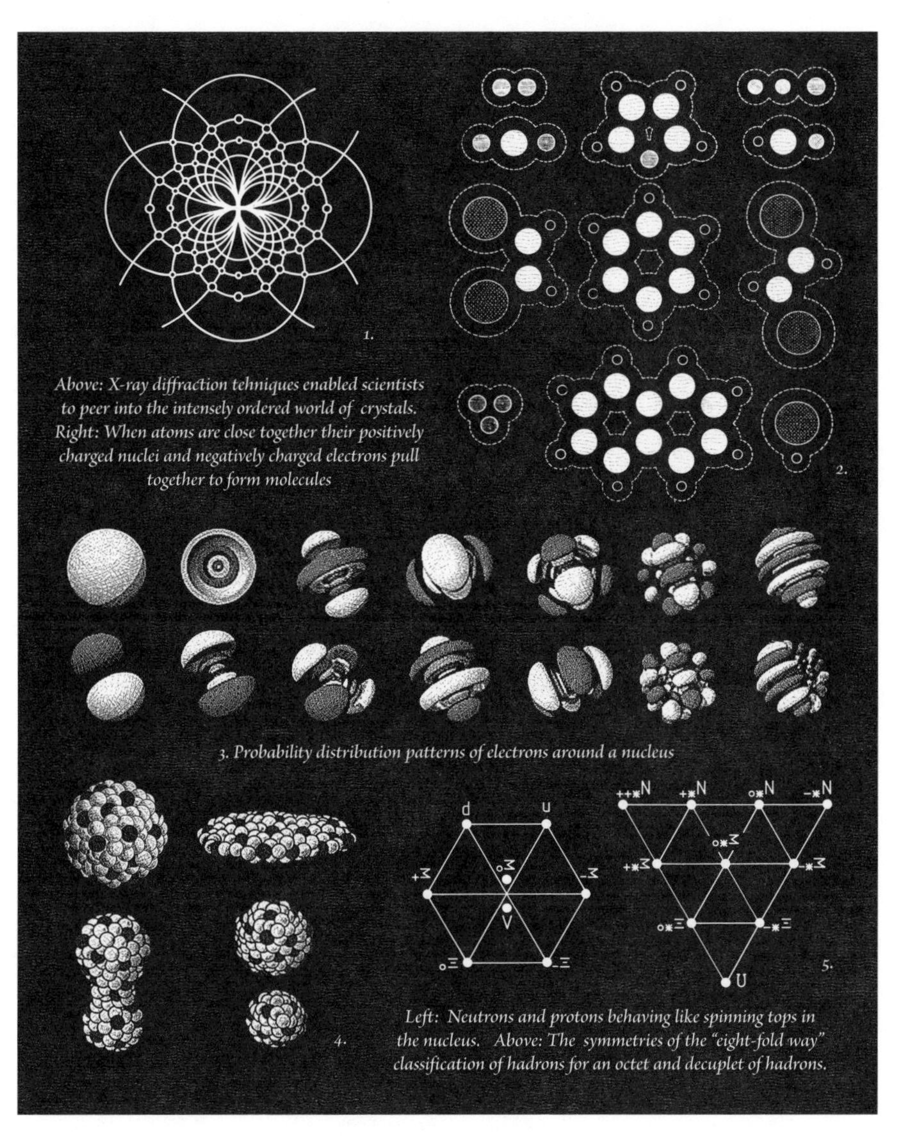

Above: X-ray diffraction tehniques enabled scientists to peer into the intensely ordered world of crystals. Right: When atoms are close together their positively charged nuclei and negatively charged electrons pull together to form molecules

3. Probability distribution patterns of electrons around a nucleus

Left: Neutrons and protons behaving like spinning tops in the nucleus. Above: The symmetries of the "eight-fold way" classification of hadrons for an octet and decuplet of hadrons.

Dorsiventrality
the symmetry of moving creatures

Animals, by definition, are multi-cellular, food-eating creatures, and practically all are capable of some form of motion; naturally, these attributes govern their general form. Whether an animal walks on the ground or burrows through it, whether it swims through water or flies through the air, its body will be made up of left and right sides that are roughly mirror versions of each other. Since they also have a front and a back (and usually a distinct top and bottom) they are not merely bilateral, but *dorsiventral*. This is the best arrangement to have if you need to move in a directed way (*examples opposite*). It is not only animals that express this symmetry; forward-moving vehicles, such as automobiles, boats, airplanes etc., are, by necessity, symmetrically disposed along similar lines.

There are other characteristics of animal dorsiventrality that developed alongside the power of locomotion. A strong forward movement obviously requires a forward-placed vision to see where one is going, and a forward-placed mouth to feed efficiently. Fins and limbs, by contrast, are best placed laterally, in symmetrically balanced positions.

Although, for the reasons given above, dorsiventrality is the abiding symmetry of the animal kingdom, it is also fairly common in the plant world—typically in zygomorphic (irregular) flowers, in the great majority of leaf forms (*below*), and in many leaf arrangements.

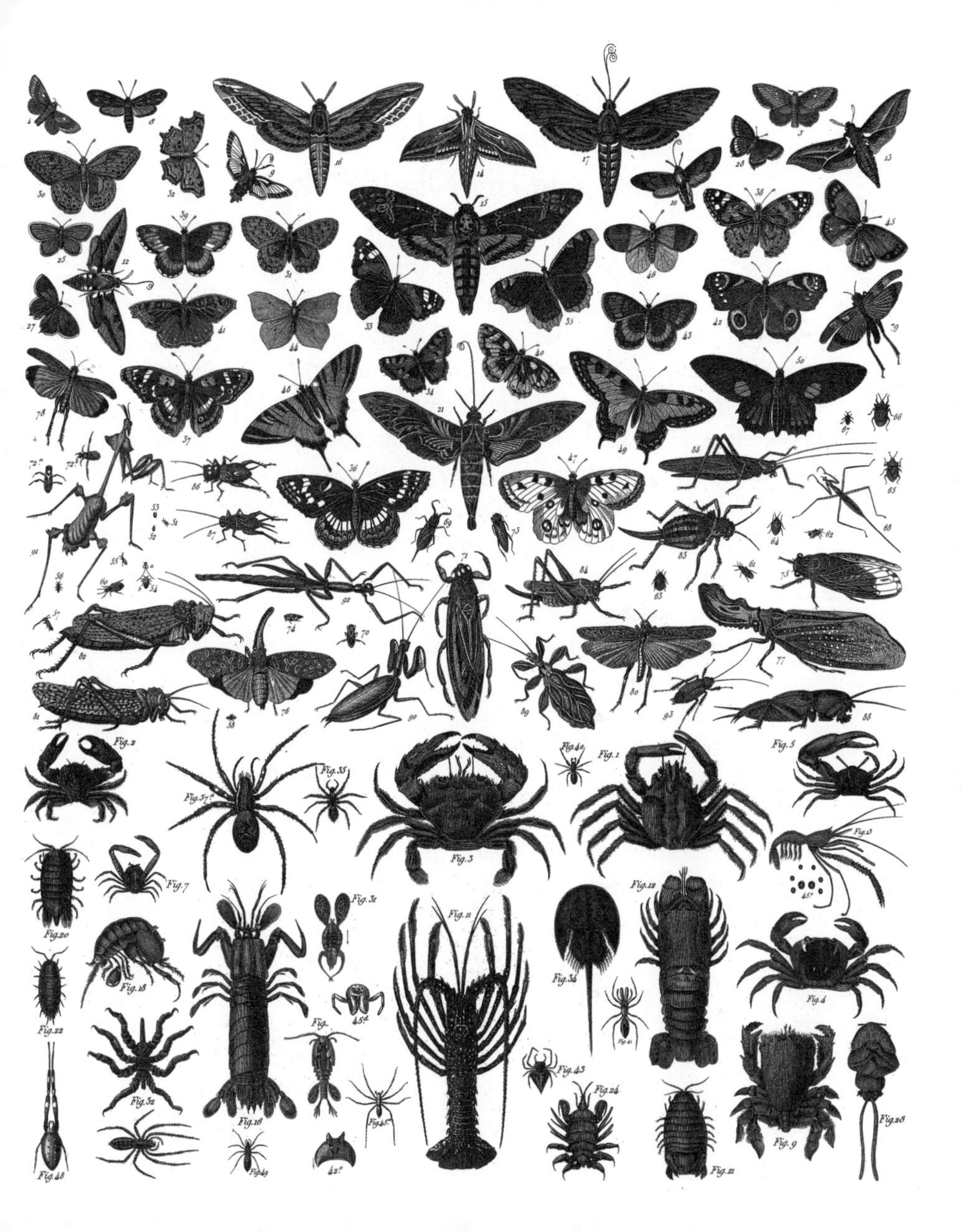

Fig.2
Fig.35
Fig.40
Fig.1
Fig.5
Fig.7
Fig.3
Fig.12
Fig.20
Fig.31
Fig.11
Fig.18
Fig.4
Fig.22
Fig.32
Fig.16
Fig.43
Fig.24
Fig.28
Fig.9
Fig.21

ENANTIOMORPHY
left- and right-handedness

Amongst other things, our dorsiventral body-form gives us a pair of hands that are similar in most respects, except that they are mirror-reversed. The same is true of our feet, of course, and of horns and butterfly wings and many other animal features (*1*). But the possibility for a figure, or an object, to exist in two distinct forms in this way is not limited to the mirror-symmetries adopted by living organisms. Any spiral, for instance, has to choose whether to go clockwise or anti-clockwise on the page (*2*), and similarly, all helices can appear in one or other of two different ways in three dimensions (*3*).

In fact the possibility of alternate forms applies to any object, animate or inanimate, that has a twist in its structure. Mollusk shells are found as both left-handed and right-handed types (some species opt for a particular handedness, in others the choice appears to be random) (*4*). There is a somewhat similar situation in the familiar twisting habits of vines and other climbing plants (the majority opt for right-handedness, but a substantial minority are lefties).

In chemistry this phenomenon is known as chirality—the most common mineral with this trait being quartz (*5*). Chirality is of particular importance in the field of organic chemistry, since many biological molecules are homochiral, that is to say, are of the same handedness, including amino acids (which are the components of proteins), and DNA (*6*). This, in effect, means that the entire chemical basis of life itself is chiral. At some early stage in the origins of life on Earth the earliest molecules to master the art of self-replication opted for a particular stereo-chemical profile, and in so doing determined the entire, right-handed, course of evolution.

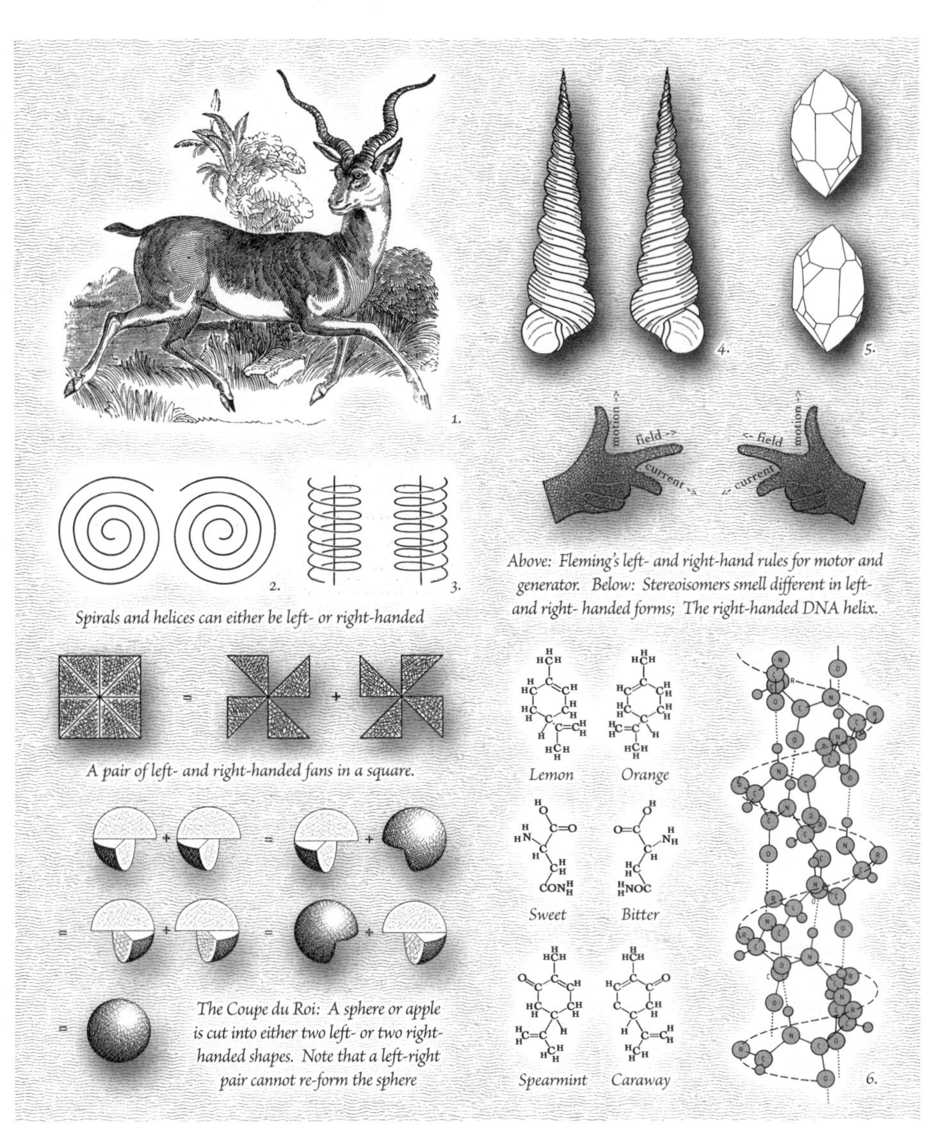

Above: Fleming's left- and right-hand rules for motor and generator. Below: Stereoisomers smell different in left- and right- handed forms; The right-handed DNA helix.

Spirals and helices can either be left- or right-handed

A pair of left- and right-handed fans in a square.

The Coupe du Roi: A sphere or apple is cut into either two left- or two right-handed shapes. Note that a left-right pair cannot re-form the sphere

CURVATURE AND FLOW

waves and vortices, parabola and ellipses

As we have considered symmetry thus far, the emphasis has been on the more static geometries of rotation, reflection, etc. With symmetries of curvature, many of which are implicated in motion and growth, these principles are extended to the dynamic (*1-3*).

The *conic sections* (*4*) were first investigated by Menaechmus in Plato's Academy in the 4th century BC, but it was not until the Renaissance that the importance of their role in physics began to be realized. In 1602 Galileo proved that the trajectory of a thrown object described a parabola. Not long after this Kepler discovered the elliptical nature of planetary motion. Later, it was realized that the hyperbolic curves could represent any relationship in which one quantity varied inversely to another (as in Boyle's Law). Discoveries of this kind epitomize the way in which a broader understanding of the symmetry principles inherent in mathematics began to uncover the hidden unity of nature.

Wave-forms also express symmetry, both in their length and period; a simple sine curve can be thought of as a projection on a plane of the path of a point moving round a circle at a uniform speed (*5*). In fact, circular motion is a component of any wave-like event. If this movement is regularly increased or diminished it produces a characteristic sine configuration.

1. Vortices formed by a split air-stream in an organ pipe

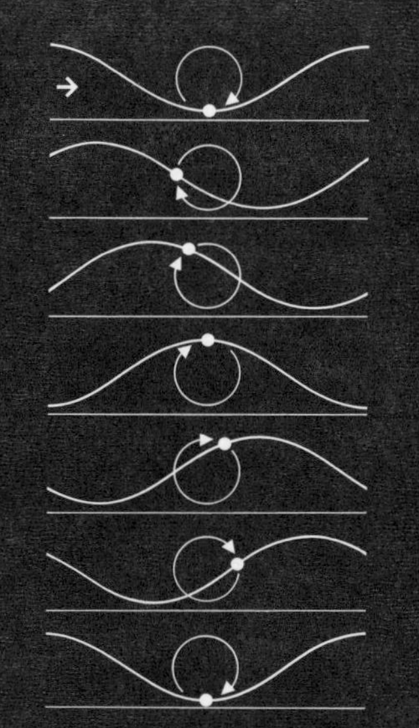

2. Wave motion in a liquid medium is essentially circular

3. A train of Karman vortices induced by an obstruction

4. Conic sections and elliptic series.

5. Upper and middle: Sine waves.
Lower: River meanders tend to adopt sine profiles.

SPIRALS AND HELICES

natures favorite structures

Of all the regular curves, *spirals* and *helices* are probably the most common. They are found throughout the natural world, in many forms, at every scale of existence—in spider-webs (*1*), galaxies (*2*) and particle tracks (*3*); in animal horns (*4*), sea-shells (*5*), plant structures, and DNA (*6*). It is clearly one of nature's favorite patterns.

In purely geometric terms, the common planar spirals are of three principle types (*below*): the Archimedean (*a*), the Logarithmic (*b*), and the Fermat (*c*). The Archimedean spiral is perhaps the simplest, consisting of a series of parallel, equidistant lines (as in old vinyl records). Logarithmic (or growth) spirals are the most intriguing and complex of all, particularly the "golden" spiral (*8*) that is associated with the Fibonacci series (*see next page*). Logarithmic spirals in general have the property of self-similarity, i.e., of looking the same at every scale. In the Fermat (or parabolic) spiral, successive whorls enclose equal increments of area, which accounts for its appearance in phyllotaxis, the arrangements of leaves and florets on a stem (and in coffee-cups).

Helices are symmetrical about an axis, so always have a particular "handedness" (*d*). Dilation symmetry can apply to helices, gradually increasing their width (*e*), and of course they may be expressed in any number of strands, in the way that ropes are laid (*f*).

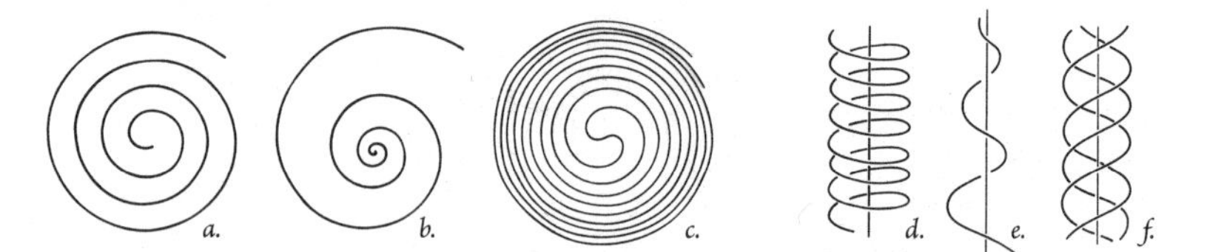

1.
2.
3.
4.
5.
6.
7. An evolute spiral.
8. The "golden" logarithmic spiral.

FABULOUS FIBONACCI

golden angles and a golden number

Around the end of the 12th century a young Italian customs officer became intrigued by (and gave his name to) a number series that has fascinated mathematicians ever since. Nicknamed "Fibonacci", Leonardo of Pisa had discovered the cumulative progression where each number is the sum of the preceding two numbers, i.e., 1, 1, 2, 3, 5, 8, 13, 21, 34 etc. He also recognized that this series has some very special mathematical properties. The Fibonacci numbers are frequently involved in plant growth patterns, notably in petal and seed arrangements. Flower petals are almost invariably fibonaccian in number; fir-cones use series of 3 and 5 (or 5 and 8) intertwined spirals; pineapples have 8 rows of scales winding one way, 13 the other way—and so on. The series is also found in phyllotaxy, the configurations of leaves and branches in plants.

The Fibonacci series is focussed on *phi*—that is to say, as they get higher the ratio between successive numbers gets closer and closer to this *golden* number. There is a related quality too in the format of successive primordia in phyllotaxy which use the "golden" angle of 137.5° (360°/phi^2). This arrangement provides the most efficient use of space in the succession of branches, leaves and flowers. Fibonacci patterns are not restricted to organic formations; they have been observed in many aspects of the physical world, from nanoparticles to black holes.

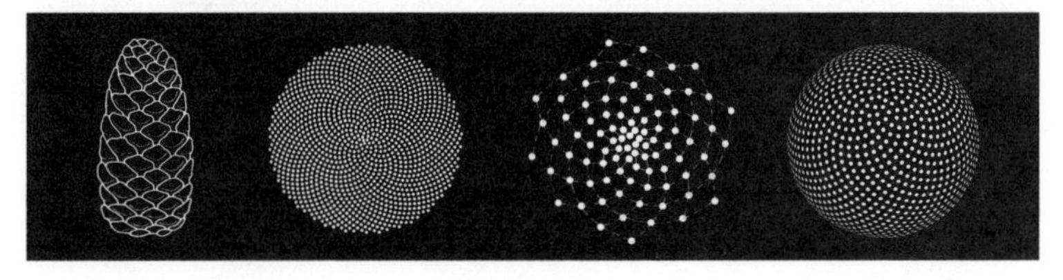

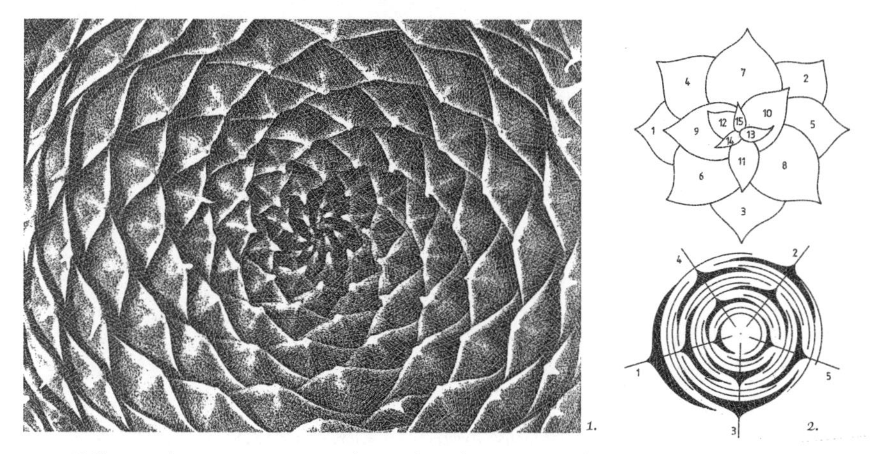

1. Phyllotaxis order 13:8 in a cactus. 2. Order 8:5, 8 leaves forming in 5 anticlockwise turns, with every 8th leaf above another 3. Another example of 8:5 phyllotaxis. 4. A rare case of Lucas phyllotaxis, order 11:7. 5. A sunflower head demonstrating 89:55 Fibonacci phyllotaxis on a Fermat spiral. Count the spirals each way.

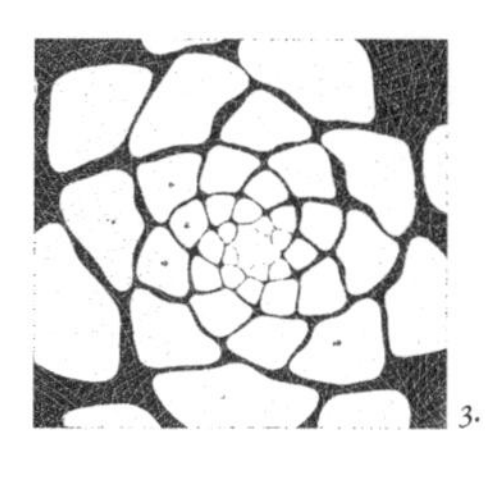

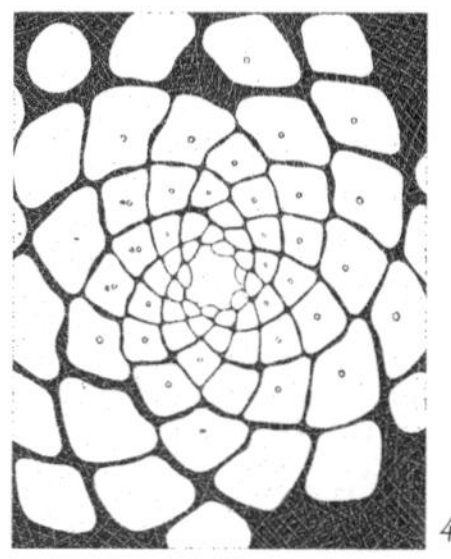

Branching Systems
patterns of distribution

Branched networks can be thought of as having a real existence, like those of trees, rivers, etc., or simply as mental concepts that exist independently of any physical representation. In the latter case, fairly complex systems can be generated from quite simple rules (*lower opposite*).

One of the more fascinating aspects of branching is that similar habits can be expressed in entirely different settings; there are, for instance, branching hierarchies in lightning strikes that closely resemble those in river systems. There may even be a close correspondence between formations that disperse and those that concentrate (*below*). In either case, functioning branching systems involve the efficient distribution of energy in one form or another—they are the simplest way to connect every part of a given area using the shortest overall distance (or least work).

The hidden symmetries operating within branching formations concern the rates and ratios of *bifurcation*. In a simple progression, for instance, three streamlets may feed into a stream, three streams into each tributary, and finally, three tributaries into a river. This sort of progression is, in fact, a common pattern, found not only in rivers and plants, but in animal vascular systems. Although the rules that determine branching in nature tend to be more involved than this, nevertheless, relatively simple algorithms may create highly complex forms.

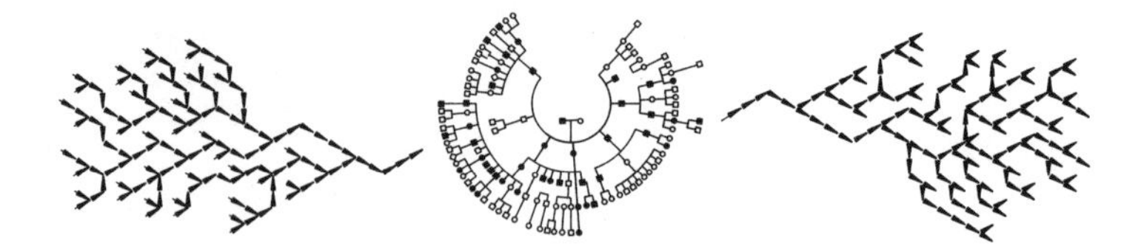

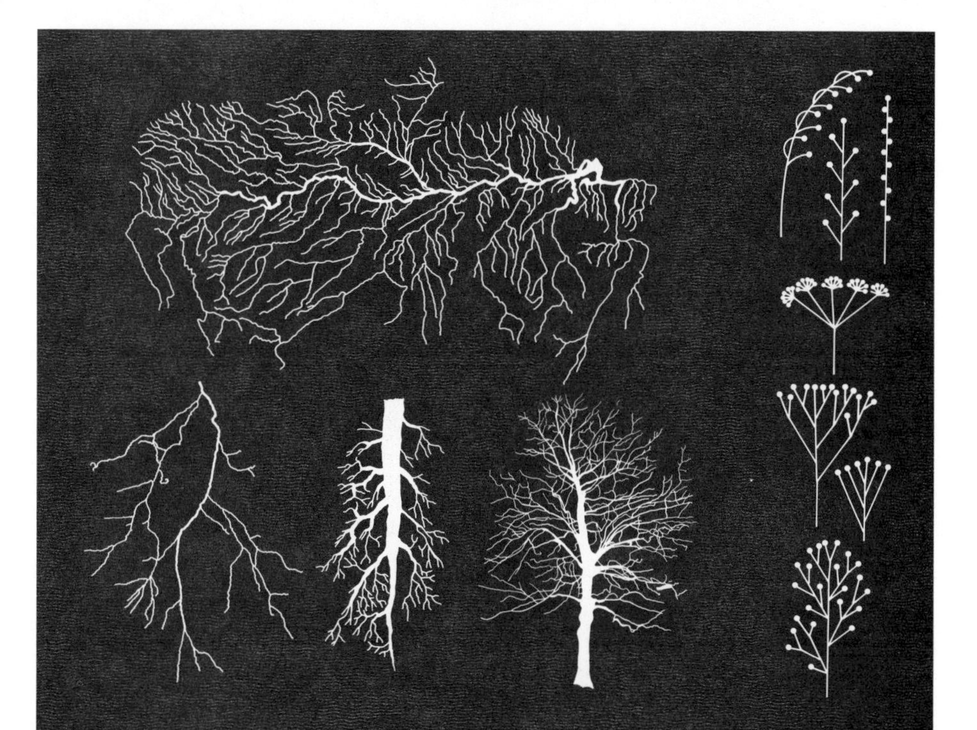

The general characteristics of all branching patterns, whether of river systems, electric discharges or those of biological systems, is that they radiate out (or converge in), and that any branches of a particular size are always outnumbered by those of the next smaller dimension.

FASCINATING FRACTALS

self-consistency to the nth degree

There are many natural phenomena, perhaps the greater part, about which the term "symmetrical" seems to have little relevance. The amorphous shapes of clouds, the rugged contours of mountains, the turbulence of streams, the patchiness of lichen, etc., especially taken together, create a distinct impression of confused irregularity. But there are consistencies in all of these things, the uncovering of which has greatly extended the notion of self-similarity, and of symmetry itself.

Many natural formations, even though they may appear highly complex and irregular, possess a recognizable statistical self-similarity. This means that they look the same across a range of different scales, and the degree of their fractality accurately measured. There is, moreover, a converse application of this notion that highly complex phenomena may have a hidden order, namely, that relatively simple formulae can create highly involved figures. The renowned Mandelbrot set (*background opposite*) is probably the best known and most complex example of this effect.

In fact, many organic structures exhibit the *fractal* properties of self-similarity; animal circulatory systems, for instance. The branching, systems of blood vessels, which repeat on an ever-reducing scale, allow the most efficient circulation of blood to every part of the body.

In mathematics many kinds of fractals are unlimited by scale and can, in theory, go on to infinity, but this is seldom the case in the real world, especially in living creatures where the rule is fitness for purpose. Blood-vessels do not reduce indefinitely, any more than the whorls-within-whorls of the fractal cauliflower extend to infinity. Nature uses fractal geometry where it is advantageous.

Fractals are linked to the enormous advances in computer science and Chaos Theory, but their geometry has a history of its own. The above forms, dating from the early 20th century, were originally seen as mathematical curiosities that demonstrated the mingling of finite spaces and infinite boundaries

Penrose Tilings & Quasicrystals

surprising five-fold symmetries

In the mid-1980s the world of crystallography was taken aback by the announcement of an entirely new kind of material, midway between the crystalline and amorphous states. What was particularly surprising about this new state of matter was that it appeared to be based on a 5-fold symmetry, apparently violating the basic laws of crystallography. Until this time the conventional understanding was that only 2-, 3-, 4- and 6-fold symmetries could create the lattice structure on which crystals were formed. The new material, Shechtmanite (*3*) (named after its discoverer), soon became classified as a quasicrystal, and other examples of these materials (which, on the scale of solids, lie somewhere between crystals proper and glass) gradually appeared. Naturally, new uses soon began to be found for these exotic materials. High-magnification microscopic images and X-ray diffraction patterns of quasicrystalline structures reveal unusual dodecahedral symmetries, and the appearance of the phi ratio.

Interestingly, the loose symmetries on which they are based had been prefigured by the Oxford mathematician Roger Penrose in the early 70s. Penrose had produced a pair of non-periodic tilings, based on approximate pentagonal symmetry (*4,5,6*). As with quasicrystals, these patterns have elements of a long-range order despite their 5-fold symmetry—and they can fill the plane in an infinite number of ways!

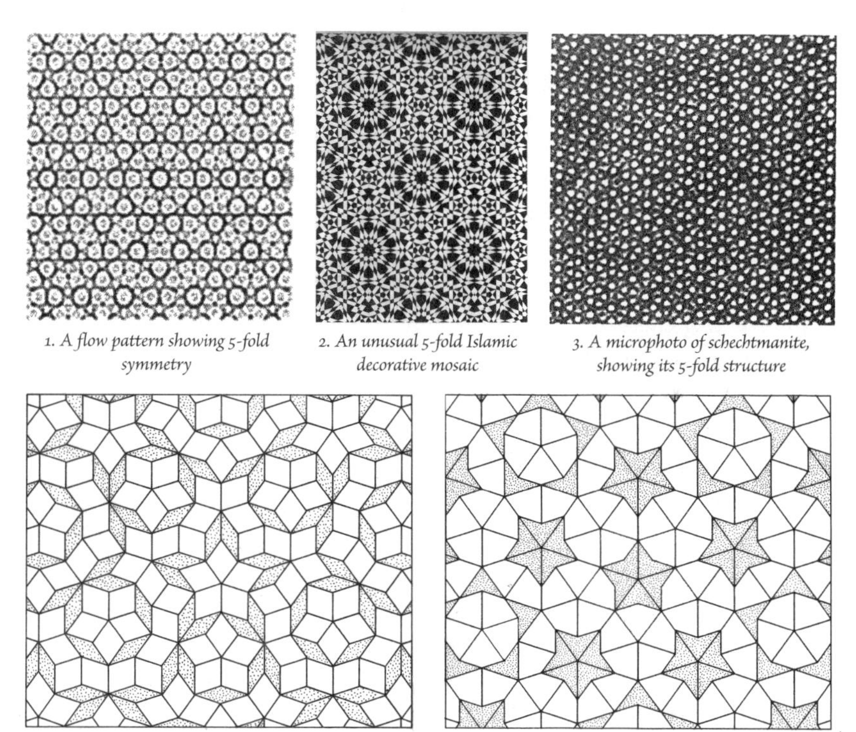

1. A flow pattern showing 5-fold symmetry

2. An unusual 5-fold Islamic decorative mosaic

3. A microphoto of schechtmanite, showing its 5-fold structure

4. Penrose tiling no. 1, using two golden diamonds

5. Penrose tiling no.2, using golden darts and arrows.

It is impossible to fill the plane using only pentagons, but Penrose tiles (6.) can be laid out in many ways

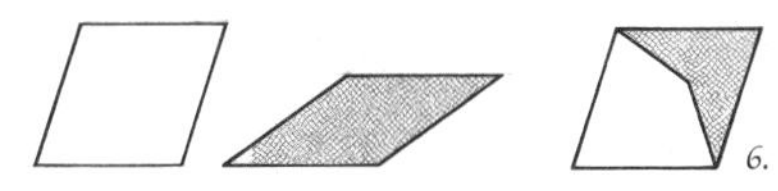

6.

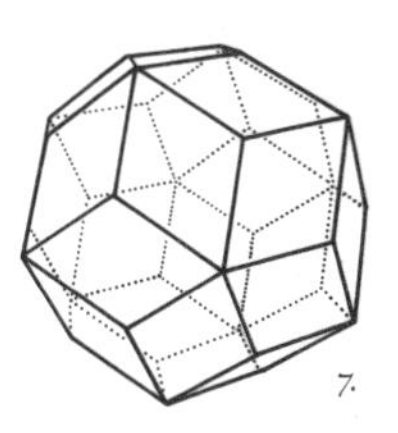

7. The rhombic triacontahedron. the 3-D analogue of a Penrose tiling, and the building block of a quasicrystal.

8. Schechtmanite "snowflakes" form when an aluminium/ manganese alloy is cooled rapidly.

Asymmetry
the paradox of inconstancy

Where does symmetry end and asymmetry begin? Take a closer look at the Roman mosaic featured on the cover of this book. Is it symmetrical or not? There is an obvious overall symmetry, but a closer examination reveals that there are different designs in each of the roundels, and as many in each of their borders. So perhaps this composition is best characterized as having a somewhat disturbed symmetry - it exemplifies the paradox mentioned in the Introduction, namely, that the notion of symmetry is essentially inextricable from that of asymmetry.

One of the most important discoveries in recent science is that the notion of "broken" symmetry has deep cosmological implications (*more of this on page 46*), but it is clear that a great many things in the world are like this. The fact is that wherever one looks there are many kinds, as well as degrees, of deviation from symmetry. The human body, for example, is bilateral (or dorsiventral) in its general form and some internal organs, like the lungs and kidneys follow this symmetry, but others, such as the alimentary canal, heart and liver do not. And even the overall symmetry is only approximate. Most of us have a dominant hand and eye, and there are subtle differences in the respective left and right sides of faces.

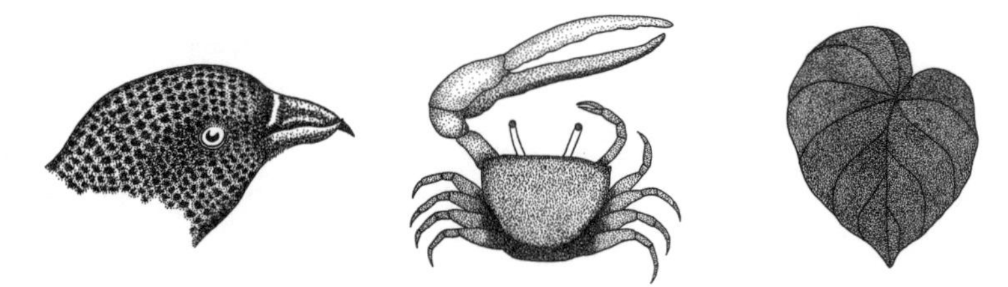

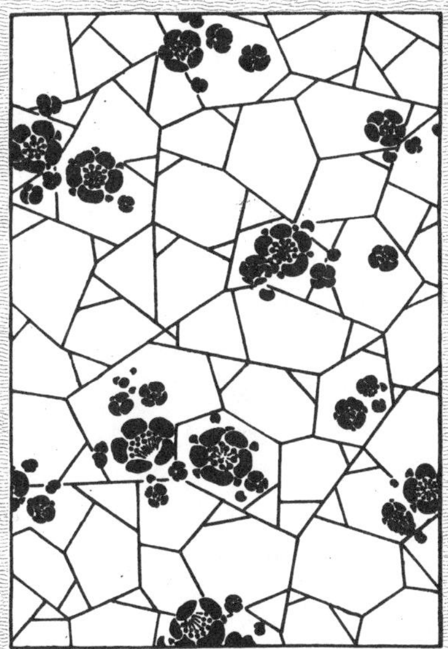

In living organisms generally the underlying reasons for the sort of deviations from bi-lateralism shown opposite derive from evolutionary fitness. Where a mirror-symmetry is appropriate or necessary it is retained, where it isn't, it may be modified or abandoned. Many species have opted for lop-sidedness in varying degrees, but we can be sure that the Crossbill, Fiddler crab and Begonia leaf each had their own very good reasons to adopt their respective asymmetries.

There is another aspect of asymmetry that should be touched on, namely that used in the field of art and design. There are various motives for deliberately introducing asymmetries into a design, these include religious or superstitious reasons, or simply the impulse to create a certain dynamic tension (this last is particularly noticeable in Japanese art). Ironically, whatever the reasons behind the use of deliberate asymmetries, there is bound to be a tacit acknowledgement of the notion of symmetry itself. This means that asymmetry in art is usually a reactive response, on some level or other, to this basic ordering principle.

SELF-ORGANIZING SYMMETRIES

regularities in non-linear systems

There are many natural patterns that present more subtle regularities than the highly ordered symmetries of crystals. Some of these are generated by quite simple rules, others by a complex of factors; many result from some form or other of self-organization. These "Li" (*opposite*) express a certain universality, their symmetries tend to be thematic and fluid rather than rigid and static. The simple ripple patterns on a sea-shore, for instance, are created by a multiplicity of contributory factors, including tides, currents and winds—not to mention the more general effects of gravity and warmth from the sun. All of these are drawn into a self-organizing, self-limiting order whose charm lies precisely in the fact that it is repetitive, yet infinitely variable.

Rivers are also self-organising. Whether they are a gentle stream or a broad torrent they tend to follow similar meandering paths. There is an invariant quality in these loops and bends that conform to well-defined mathematical parameters. Similar constraints govern the hierarchical patterns of river-drainage. Rivers shape the terrain that they flow through, and are in turn shaped by it, but there are many subtle factors that limit and influence their form.

"Scale-invariant" symmetries also appear in fracture patterns of the kind found in mud cracks and ceramic crackle-glaze. Formations of this sort usually appear as a result of stresses induced by shrinkage. There are variations in the modes of cracking in different materials and in different conditions, but all are characterized by an overall consistency, and many have scaling properties. They are formed, and limited, by the release of stress, so they are progressive and self-organizing—and of course they tend to be fractalline in nature.

SYMMETRIES IN CHAOS

regularities in highly complex systems

Invariance equates with symmetry, so on the face of it, turbulence, which is the very image of a totally disturbed system, would appear to be an unlikely candidate for symmetries of any kind. The physics of turbulent systems was for a long time one of science's most intractable problems, it is still not completely understood, but the recognition of the role of *strange attractors* in the process has brought new insights and a new mathematical instrument to bear on such complex systems.

The cryptic geometry of strange attractors was part of the new non-linear maths of *Chaos theories* (the revolution in which fractals first appeared). It involves the concept of viewing dynamical systems as occupying geometrical space, the coordinates of which are derived from the systems variables. In linear systems the geometry within this phase space is simple, a point or a regular curve; in non-linear systems it involves far more complex shapes, the "strange" attractors. One of the most famous of these is the Lorenz attractor (*1,2*), which forms the basis of chaotic models of weather prediction (including Ice-Ages). Another classic example is the "dripping tap experiment" (*3*) where beautiful regular forms are found within apparent randomness.

As we have seen, Fractal geometry is intrinsic to many aspects of Chaos theory—and fractals are, predictably, firmly associated with attractors. In fact all strange attractors are fractal, as is *Feigenbaum mapping*, which is a sort of master attractor. The *Feigenbaum number* which lies at the heart of this mapping, predicts the complex, period-doubling values across a whole range of non-linear phenomena, including turbulence (*4*). The Feigenbaum value is recursive, and appears whenever there is repeated period doubling. It is, in short, a universal constant like *pi* or *phi*, and has a similar symmetrical potency.

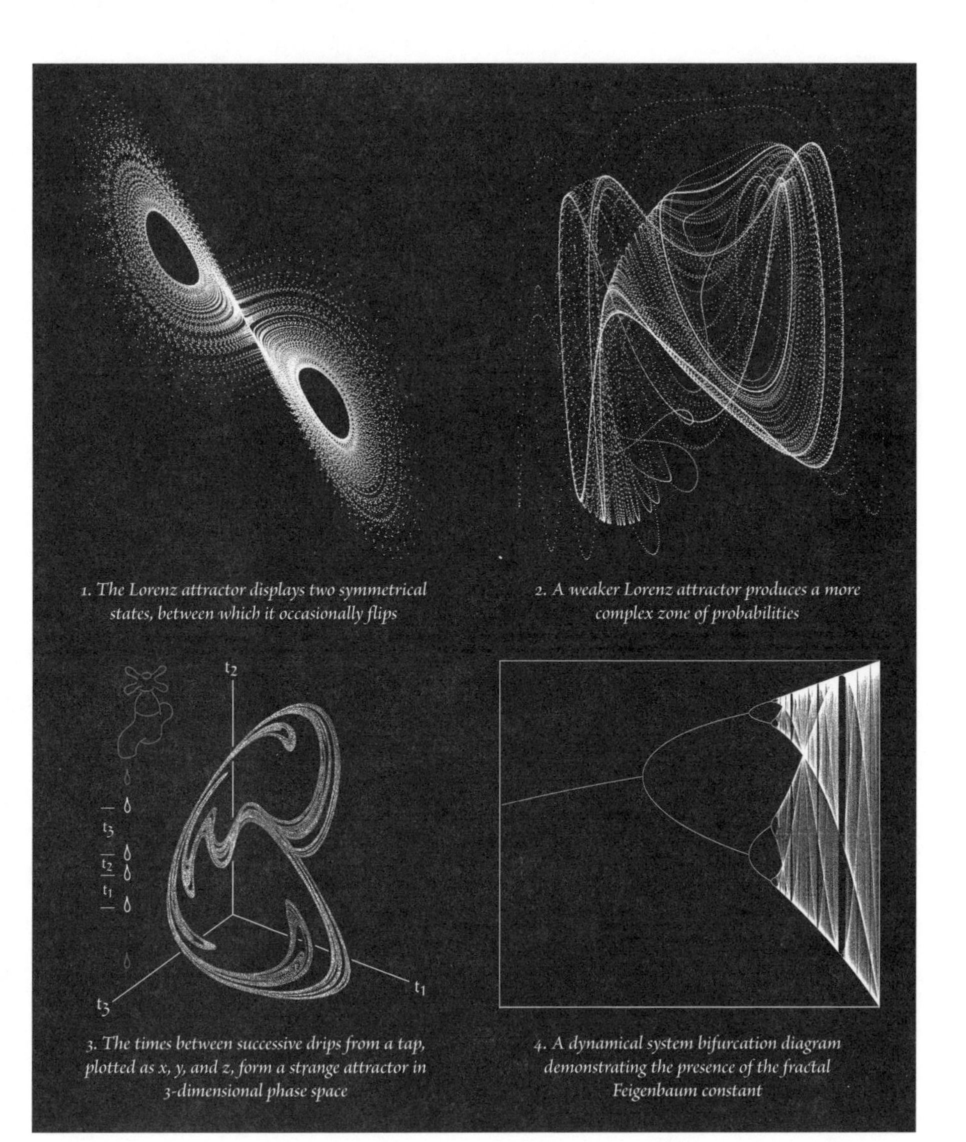

1. The Lorenz attractor displays two symmetrical states, between which it occasionally flips

2. A weaker Lorenz attractor produces a more complex zone of probabilities

3. The times between successive drips from a tap, plotted as x, y, and z, form a strange attractor in 3-dimensional phase space

4. A dynamical system bifurcation diagram demonstrating the presence of the fractal Feigenbaum constant

Symmetry in Physics

invariance and the laws of nature

Since the amount of energy in a closed system is invariant, the law of conservation of energy is now seen as a symmetry law. In fact there is a real sense in which the history of physics (at least in the modern period) might be characterized as a successive uncovering of such universal conservation principles. The great discoveries of Galileo and Newton concerning gravity, for instance, were essentially the recognition of physical laws that deeply affect the material world, and yet are in some sense independent of it. Newton's Law, in postulating a symmetrical force acting on all objects, discovered the invariant quality of gravity, i.e., that it is the same everywhere in the universe. By extending these laws to a moving, even an accelerating observer, Einstein added further symmetries. This was the basis of his theory of General Relativity.

Gravity is now recognized as just one of four fundamental forces of nature underlying all natural phenomena. In one of the greatest intellectual achievements of the 20th century the mathematician Emmy Noether established the connection between these dynamic forces and the abstract notion of symmetry. Since the laws of physics apply equally in every part of ordinary space they are regarded as possessing translational symmetry, which, on the most basic level, is a consequence of (or equivalent to) the law of the Conservation of Momentum. Physical laws do not change over time either, which means that they are symmetric under translations in time, leading to a *conservation law*, in this case, the Conservation of Energy. In physics, there is now an absolute connection between symmetry and the laws of nature, so that physicists consciously search for *invariance* in their quest for new conservation laws. Reality, it seems, is threaded through with concealed symmetries.

Left: Emmy Noether and her 1915 theorem: "For every continuous symmetry of the laws of physics, there must exist a conservation law. For every conservation law, there must exist a continuous symmetry."

Below: The refraction or bending of light in different media is most simply understood in terms of Noether's Theorem by realising that a photon always takes the quickest path available from source to destination.

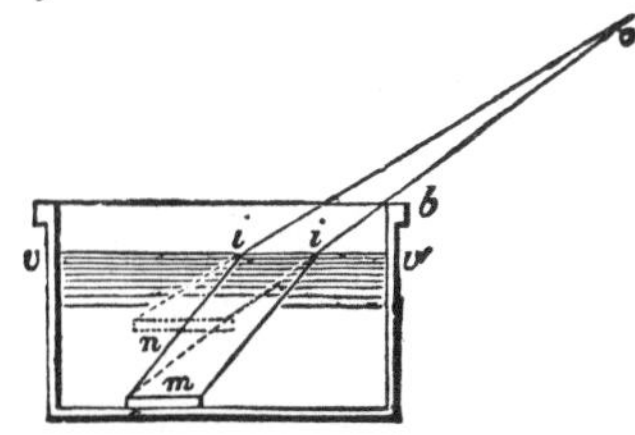

Symmetry in Art

constraint and creative potentiality

The artistic impulse appears to be a basic human response, but its aims, methods and roles within societies are as diverse as the cultural settings themselves. Art can have a magical or religious purpose, and may be representational or decorative—but whatever its aims or functions, expresses a style which ties it to a particular time and place. Where symmetries of any kind are present in art they will be intimately involved with the particulars of a style, since symmetry, in art as elsewhere, is an organizing principle. It would seem that humans are symmetry-conscious creatures; we are pattern-seekers by nature, so symmetry principles are never entirely absent as a consideration in art generally. The role of ratio, proportion and symbolism in the Fine arts and architecture is examined later (*see page 50*), but broadly speaking, it is in the decorative arts where symmetrical arrangements are most in evidence.

The arts of tribal people just about everywhere use the basic symmetry functions of reflection and rotation. Bi-lateral arrangements in particular are an effective way of organizing a composition, a method widely used in both "primitive" and advanced societies. Dihedral symmetries are also widespread, finding their ultimate expression in the fine rose-windows of Gothic cathedrals (*10*). There are, however, great cultural variations in the role of symmetry in art. In some it plays a small part, while others thoroughly explore its possibilities. Interestingly, this fascination (or the lack of it), applies across a whole range of societies, from tribal to those more advanced—right up to the present in fact. Naturally, those artistic traditions whose taste inclined toward symmetry have always tended to develop a richer vocabulary in this respect, and have explored a greater range of decorative possibilities.

1. *Pueblo pottery* 2. *Celtic strainer* 3. *Inca plate*

4. *Islamic motif*

5. *Seljuk mosaic*

6. *Romanesque device*

7. *Persian ceramic*

8. *Box; Nth. Pacific coast*

9. *Detail of Ainu coat*

10.

A Passion for Pattern

the perennial appeal of repeating designs

Pattern arises almost of itself from any repeated operation (such as knitting, weaving, brickwork, tilework, etc.), but patterning has often become an integral part of a culture's stylistic conventions in its own right. In fact, although most cultures have used pattern as part of their decorative repertoire, some, at different periods and in different parts of the world, seem to have become positively fixated on patterning as a mode of artistic expression. The complex varieties of Islamic pattern are well-known, but there have been equally strong traditions in the Celtic world, in Meso-america, and in Byzantium, Japan and Indonesia. Even those of us from cultures that are not so pattern-obsessed are perfectly capable of appreciating repeated ornament. It has a certain universality.

Regular patterning always involves a measuring of the space to be decorated. Because of this, the artist, knowingly or otherwise, engages with the rules governing the symmetry-groups of plane division (*see Appendix, page 56*). In practice, these limitations are not so much a constraint on design as a further opportunity to introduce variety.

Interestingly, at least two artistic traditions, those of Ancient Egypt and Islam, came fairly close to using all 17 classes of planar patterns. The unconscious, but systematic exploration of symmetry groups in this way would seem to blur the distinction between the artistic activity of pattern creation and that of science, whose entire enterprise could be characterized as pattern detection.

SYMMETRIA
sublime proportions

The Renaissance saw a revival of interest in classical notions of symmetry. The idea of symmetry as a harmonious arrangement of parts that was propounded by Vitruvius, a Roman, actually derived from older, Greek, views of a fundamental order and harmony within the universe. This strand of thought is generally associated with the influential philosophy of Pythagoras and his followers, for whom geometry (and in particular the geometry of ratios and proportion) was the key to a deeper understanding of the cosmos.

The idea of a harmonious correspondence between the parts of a system and the whole is a compelling one—and there is a great deal of evidence that certain special proportions were employed in ancient architecture, both in the European and other traditions. This usage was continued to some extent in those cultures that inherited the classical tradition—in the Islamic world, and in Gothic cathedrals for instance, as well as in the Renaissance revival.

In his seminal work *De Architectura*, Vitruvius made the definitive statement on these principles—"Symmetry results from proportion; proportion is the commensuration of the various constituent parts with the whole." Under the influence of these ideas, the Renaissance architect Alberti introduced a Pythagorean system of ratios into architecture, relating these concepts to dimensions of the human body —an idea that was enthusiastically taken up by the artists Albrecht Dürer and Leonardo Da Vinci, among others.

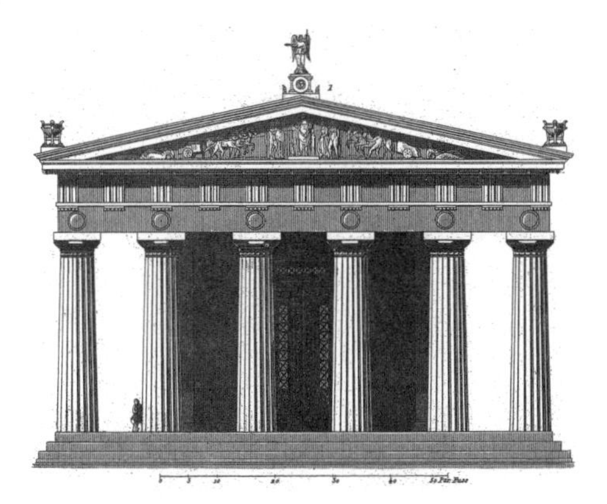

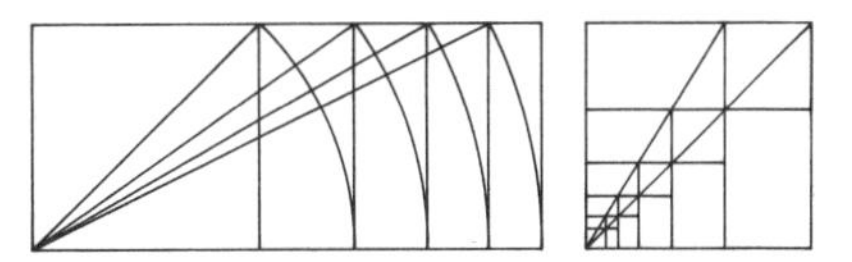

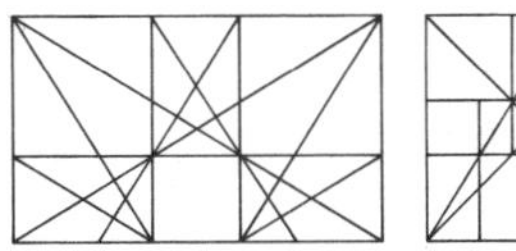

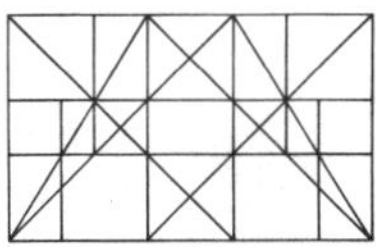

1. Modular series of proportionate rectangles can be generated from various ratios, including root 2, root 3, and phi

2. Many ancient cultures used systems of harmonious proportion in their architecture

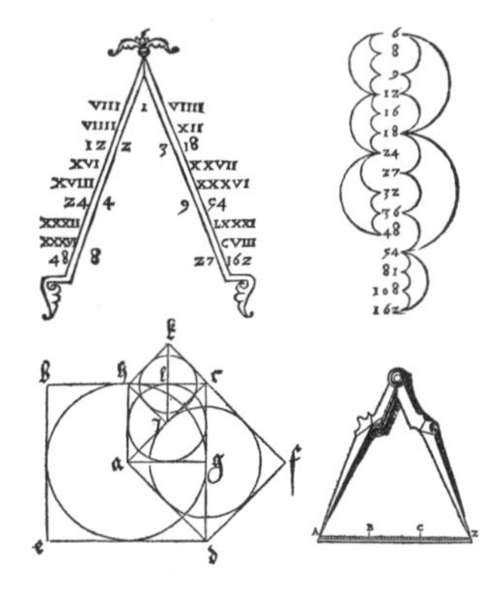

FORMALISM
symmetry symbolizing stability

Symmetry is frequently involved in places and occasions that seek to project the notion of formality, which itself is necessarily bound up with concepts of the status quo and, by extension, of social order and constitutional rule. This is the underlying reason for the symmetries in the architecture of palaces, governmental buildings and places of worship. Ceremonial displays, formal gardens and formal dancing are also based on regular arrangements for similar reasons. Symmetry is used here to symbolize qualities of endurance and stability—which of course any established order would wish to identify with (and which its followers would wish to imitate). The tacit intention of formalism then, in any sphere, is an alignment with some or other perceived notion of order.

In any formal scheme of this kind individualism tends to be submerged within the greater pattern. The so-called Ancient Mature civilizations, (such as those of Pharaonic Egypt, Mesopotamia and Meso-America) in which all behavior was highly prescribed, provide the most extreme examples of formalized societies. The massive monuments that they left offer the most compelling evidence of their rigid world views. The awesome symmetries of pyramids, ziggurats and the like were not only the link between heaven and earth, but were models of the intensely hierarchical societies that produced them. Above all, their impressive, symmetrical monuments symbolized enduring stability.

These ancient civilizations declined under the impact of more dynamic societies, but their use of symmetry as a metaphor for official order and decorum persisted. Ritual and ceremonial still have an important role in political life, and symmetry is still an important part of the whole symbolism of legitimacy.

PORTRAIT DES CHASTEAVX ROYAVX DE SAINCT GERMAIN EN LAYE
A La riuiere de seyne
B Le iardin des canaulx a 117 toises de long et 55 de large
C Le iardin en pente a 117 toises de long et 52 de large
D Le iardin qui est entre les deux bosquetz a 74 toises de long et 40 de large
E Les deux bosquetz ont chascun 30 toises de long et 40 de large
H la grotte de neptune
I la grotte des orgues
K Les chapelles du roy et de la royne
L Les deux petitz iardins ont 6 toises de long et 23 de large
M La grande salle a 14 toises de long et 8 de large
N Les galleries du roy et de la royne ont 34 toises de long et 3 de large
O La cour de deuant la grande salle a 30 en carre
P Les deux cours des officiers ont 28 toises de long et 20 de large
Q La cour dentre les deux chasteaux a 134 toises de long et 74 de large
R Le ieu de paulme.
T La basse cour a 79 toises de long et 26 de large
V Le grand iardin du roy a 150 toises de long et 97 de large
Du depuis la cour qui est deuant la grande salle iusques a la riuiere il i a 28 toises de pente
Le parc a 1991 toises de long et 312 de large

EXPERIENTIAL SYMMETRIES

percepts and precepts

It is clear that symmetry is an all-encompassing principle. We have seen that it is involved in natural structures in countless ways, and that symmetry concepts have become an essential tool for a deeper understanding of the physical world. It is also apparent that symmetry has an aesthetic dimension, and that it contributes to that most elusive of concepts, Beauty. What is rather less tangible is the part that this ordering principle plays in our ordinary experience of life as social beings—needless to say that it has an important role here too. To begin with, symmetry is an essential component of the basic social norms of reciprocity. We expect fair dealings in social exchanges, and this basic sense of fairness is as natural to humans as it apparently is to our cousins, the higher primates. By extension, any system of justice is bound to reflect these notions of proportionality; this is symbolized by the image of the balance-scale, that most graphic representation of symmetry.

Notions of proportionality and reciprocity also make an essential contribution to every system of religious belief. Most religions hold that our actions in our present lives will determines our fate in the hereafter, to an exact degree. Heavens usually have their inverted equivalent in the form of Hells. Not all religious injunctions are so oppressive, though . . .

Perhaps the most elegant of all religious precepts comes in the form of the Golden Rule, which was promulgated by many great spiritual leaders, including Confucius, Jesus Christ and Hillel (it is also found in *The Mahabharata* and Leviticus, and recommended by the Stoic philosophers). The Rule recommends that we treat others as we ourselves would wish to be treated, an ethical stance that is hard to improve upon—and one that expresses a beautiful symmetry.

Above: A kaleidoscope turns a random group into a beautiful object.

Below: Matter and antimatter, an electron and a positron.

Above: The jostling action of quanta assume an overall symmetrical distribution.

Below: Symmetry in a coffee cup.

Appendix - Groups

POINT-GROUPS:
2-D symmetry about a center, with rotation around a center (left); reflection about a line (middle); and reflection plus rotation (right).

LINE-GROUPS:
2-D symmetry along a line. The combination of the operations of repetition, rotation and reflection about a line produce seven line groups that may, in theory, extend to infinity (right).

NETS:
The five basic nets (below) are the grids on which the variations of plane patterns are constructed.

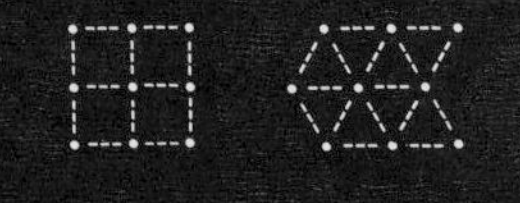

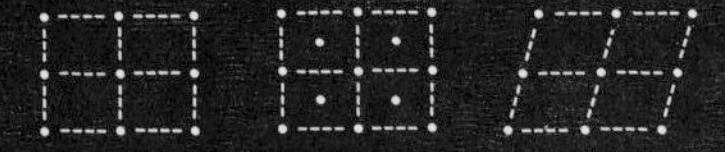

PLANE GROUPS:
In creating plane patterns from a given motif we encounter a similar set of rules (which similarly lead to a range of creative possibilities). Using the basic nets, a motif may be moved through every combination of rotation and reflection to produce precisely seventeen configurations (below).

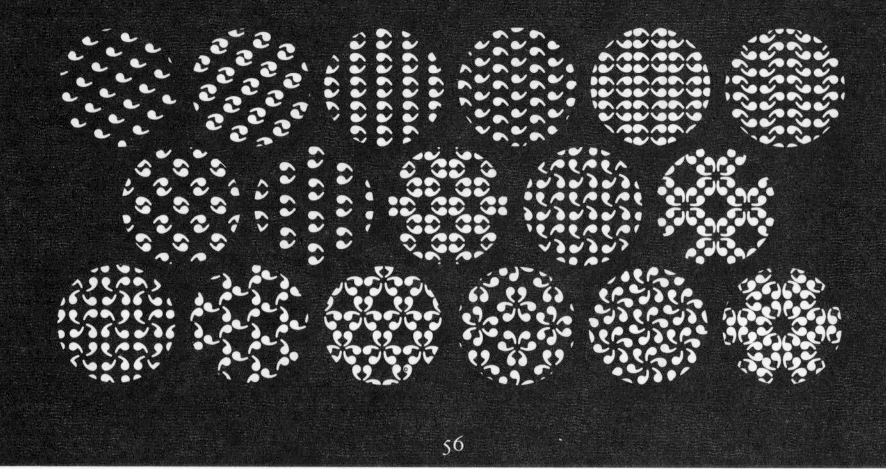

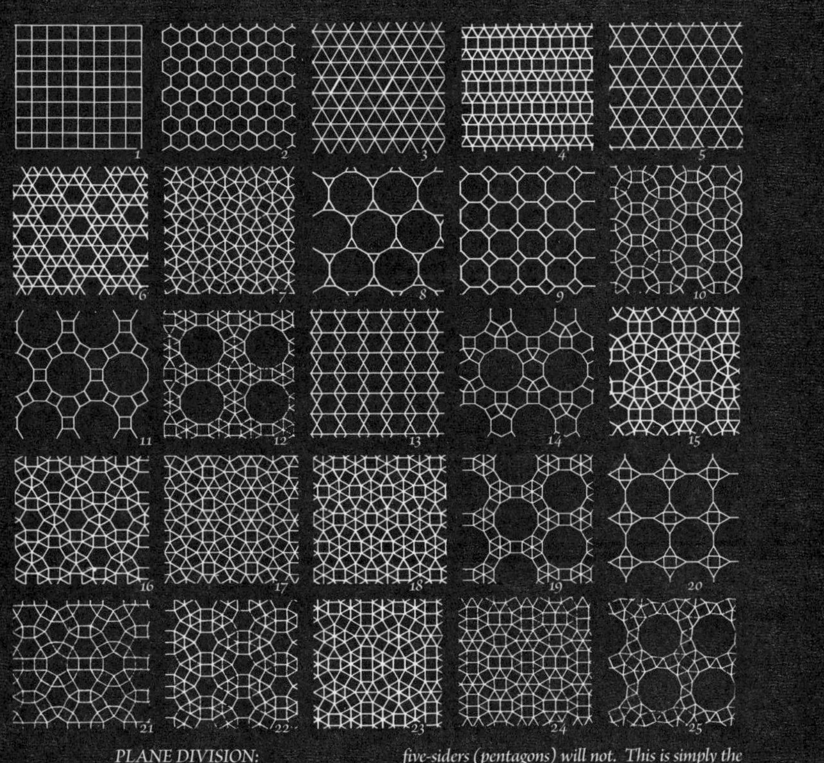

PLANE DIVISION:

Similar constraints govern the regular division/tiling of the plane. There are only three ways to do this using regular polygons. Those with three, four and six sides (the square, equilateral triangle & hexagon) will fill the plane by themselves, but five-siders (pentagons) will not. This is simply the peak of a hierarchy of plane-division classification. As well as the three regular divisions (1-3), there are eight semi-regular grids (4-11), and fourteen demi-regular grids (12-25), which together make up all the variations using regular polygons.

GLOSSARY

Algorithm *A mathematical rule for computation involving a succession of procedural steps*

Array *In maths, a regular matrix*

Attractor *In dynamical systems, a set to which a system tends to evolve*

Basic Nets *In plane-division this is the framework that creates the unit-cell which provides the mode of repetition*

Bilateral *Commonly, having two equal but reversed sides; technically, reflected about a mirror-line in two dimensions; or about a mirror-plane in three dimensions*

Bifurcation *The process of division into two branches*

Chaos theory *Mathematical theories dealing with apparent randomness deriving from precise, deterministic causes, and hidden consistencies in complex, non-linear dynamical systems*

Chiral *Of a shape that is not superimposable on its mirror-image*

Congruence *In geometric symmetry this means that the separate elements involved in the symmetry correspond in every detail, and that the distance between any two points on any part of these elements is regular*

Conservation Law *A law which states that the total value of a given quantity is not changed in any reaction*

Continuous *Term applied to symmetry groups with an infinite number of symmetry operations, i.e., those of a circle*

Conic sections *Curves of the second degree (since they intersect with a straight line at only two points).*

Curves *Can be thought of as a point that moves along a continuous path, or locus; they are symmetrical when the direction of this locus is self-consistent.*

Dihedral *Finite, centered arrangement around mirror-lines*

Dilation *Symmetrical transformation achieved by enlargement (or reduction) by means of lines radiating from a center*

Discrete *Term applied to symmetry groups that involve discrete steps, and have no infinitesimal operations, as in an equilateral triangle*

Dorsiventral *Reflection about a single mirror-plane in three dimensions*

Feigenbaum mapping *Self-similar mapping that is unchanged by "renormalisation", i.e. has constant scaling ratio*

Feigenbaum number *A mathematical constant, 4.6692016; symbolized by d; the ratio between successive periodic doublings in Feigenbaum mapping*

Fractal *A geometrical figure that is recursive and scaling, i.e., that repeats itself on an ever reduced scale*

Gnomon *A geometrical shape which, when added to or subtracted from another, results in a figure similar to the original*

Golden Section, Golden Ratio *That division of a line that leaves the ratio of the smaller segment to the larger equal to that of the larger segment to the whole line*

Group Theory *The mathematical language of symmetry (see note below)*

Invariance *A constancy; in maths, an expression or quantity that is unchanged by a particular procedure; in physics, an equality of laws in space or time, virtually synonymous with symmetry*

Isometry *Any movement or transformation that maps a figure onto a congruent figure. It is qualified by being either direct or opposite*

Isomorphic *Having the same abstract structure, even when described in different terms*

Movement *The change of a congruent object from one symmetrical position to another; may be direct or opposite*

Periodicity *The regular spacing of elements in symmetries*

Phase transition *A critical transition of a system from one state to another, usually associated with a change in symmetry, i.e., melting, boiling, magnetism*

Phi *The "golden" number, $(\sqrt{5}-1)/2 = 0.6180339887$; symbolized by ϕ; it can be squared by adding 1, and its reciprocal found by subtracting 1*

Point symmetries *Symmetries around a point or line*

Reflexion *An indirect or opposite isometric movement about a mirror-line in two dimensions; or around a mirror-plane in three dimensions*

Rotation *An isometric movement around a point; the element of symmetry can be rotated 2, 3, 4 or more positions*

Spirals and helices *Are symmetrical by virtue of the regularity with which they wind around a center or axial line respectively*

Strange attractors *A chaotic attractor, or one that has non-integer dimensions; see Attractor*

Symmetry groups *The group of all isometries, under which it is invariant with composition as the operation*

Tortuous curves *Regular curves in 3-D; measured by their changes of direction across three continuous points*

Transformation *A rule for a movement in a symmetry*

Translation *Transformations that slide objects along without rotating them*

Wave equation *A differential equation which describes the passage of harmonic waves through a medium. The form of the equation depends on the nature of the medium and on the processes by which the wave is transmitted*

A NOTE ON GROUP THEORY

One of the most remarkable aspects of symmetry groups is the extent to which they are represented in nature - indeed at the fundamental level nature could actually be said to be defined by symmetry. Group theory, the basic mathematical description of symmetry, classifies the various types according to the operations involved, i.e., rotations, reflections, repetitions, and the various combinations of these. The principle division in this general scheme is between discrete and continuous symmetries. Remembering that a symmetry is defined by the "movement" required to restore an object to its original position, a ***discrete*** *symmetry relies on a series of discrete steps to achieve this, as in the regularities of, say, an equilateral triangle. The point-groups and lattice groups that have been variously encountered in this book belong in this category.* ***Continuous*** *symmetries, by contrast, are constant over infinitesimal movements of angle and distance; circles and spheres, in 2-D and 3-D respectively, are of this kind. Continuous symmetry groups are mathematically described by a particularly elegant branch of algebra known as Lie Group Theory.*

In the late 19th century the French mathematician Elie Cartan (1869-1951) used Lie groups to classify every possible variant in this class of symmetry. This exhaustive work remained as a somewhat obscure branch of mathematics until the early 1960s, when it was recognized by Murray Gell-Man of Caltech as the perfect instrument to deal with the plethora of sub-atomic particles that were being discovered at that time. It soon became apparent that the Lie symmetries SU(3) fitted perfectly with the emerging field theories of quantum physics, to the extent that the existence and properties of some particles were predicted prior to their actual discovery. The four basic forces (gravitational, electromagnetic, and the weak and strong nuclear) are similarly described, using the gauge symmetry groups U(1) X SU(2) X SU(3). This means that present cosmological views, centered around the so-called Standard Model of fundamental particles and anti-particles, are conceived entirely in terms of symmetry groups. The challenge for contemporary cosmologists is to "unify" the basic forces with this new periodic table, in one grand symmetrical scheme.